Library of
Davidson College

BASIC Programs for Chemical Engineering Design

James H. Weber
Department of Chemical Engineering
The University of Nebraska
Lincoln, Nebraska

MARCEL DEKKER, INC.   New York and Basel

Library of Congress Cataloging in Publication Data

Weber, James H., [date]
   BASIC programs for chemical engineering design.

   (Chemical industries ; v. 16)
   Includes index.
   1. Chemical engineering--Computer programs.
   2. Basic (Computer program language)   I. Title.
   II. Title: B.A.S.I.C. programs for chemical engineer-
   ing design.
   TP184.W43 1984   660'.028'5425   84-7612
   ISBN 0-8247-7138-9

COPYRIGHT © 1984 by MARCEL DEKKER, INC.   ALL RIGHTS RESERVED

Neither this book nor any part may be reproduced or transmitted
in any form or by any means, electronic or mechanical, including
photocopying, microfilming, and recording, or by any information
storage and retrieval system, without permission in writing from
the publisher.

MARCEL DEKKER, INC.
270 Madison Avenue, New York, New York 10016

Current printing (last digit):
10 9 8 7 6 5 4 3 2 1

PRINTED IN THE UNITED STATES OF AMERICA

CHEMICAL INDUSTRIES

A Series of Reference Books and Text Books

Consulting Editor
HEINZ HEINEMANN
*Heinz Heinemann, Inc.,
Berkeley, California*

Volume 1: Fluid Catalytic Cracking with Zeolite Catalysts,
Paul B. Venuto and E. Thomas Habib, Jr.

Volume 2: Ethylene: Keystone to the Petrochemical Industry,
Ludwig Kniel, Olaf Winter, and Karl Stork

Volume 3: The Chemistry and Technology of Petroleum,
James G. Speight

Volume 4: The Desulfurization of Heavy Oils and Residua,
James G. Speight

Volume 5: Catalysis of Organic Reactions,
edited by William R. Moser

Volume 6: Acetylene-Based Chemicals from Coal and Other Natural Resources, *Robert J. Tedeschi*

Volume 7: Chemically Resistant Masonry,
Walter Lee Sheppard, Jr.

Volume 8: Compressors and Expanders: Selection and Application for the Process Industry, *Heinz P. Bloch, Joseph A. Cameron, Frank M. Danowski, Jr., Ralph James, Jr., Judson S. Swearingen, and Marilyn E. Weightman*

Volume 9: Metering Pumps: Selection and Application, *James P. Poynton*

Volume 10: Hydrocarbons from Methanol, *Clarence D. Chang*

Volume 11: Foam Flotation: Theory and Applications, *Ann N. Clarke and David J. Wilson*

Volume 12: The Chemistry and Technology of Coal, *James G. Speight*

Volume 13: Pneumatic and Hydraulic Conveying of Solids, *O. A. Williams*

Volume 14: Catalyst Manufacture: Laboratory and Commercial Preparations, *Alvin B. Stiles*

Volume 15: Characterization of Heterogeneous Catalysts, *edited by Francis Delannay*

Volume 16: BASIC Programs for Chemical Engineering Design, *James H. Weber*

Volume 17: Catalyst Poisoning, *Louis L. Hegedus and Robert W. McCabe*

Additional Volumes in Preparation

Preface

To say that computers have revolutionized technical calculations is to state the obvious. With computers many problems can now be quickly solved which hitherto were regarded as too tedious to undertake, or answers were estimated after making a few rough calculations. Computers have changed every aspect of engineering and science from education to the most complicated applications.

The advent of the pocket computer means that the engineer and scientist have available while they are in the field or laboratory the means of carrying out many types of calculations. And if the pocket computer is equipped with a printer, results may be obtained in hard copy.

It is with these thoughts in mind that the programs presented in this work were developed. The TRS-80, pocket version, was selected because of its use of the BASIC language and its printing and recording capabilities. With this, or a comparable computer, technical people have powerful computing tools available to them regardless of location.

The programs offered fall into two broad categories. First, there are a number for the calculation of physical and thermodynamic properties. Programs are given for equations of state, correlations for the prediction of transport properties, calculations of thermodynamic data, and other items. In the second category are programs which are used in the design of equipment. These include the sizing of orifices, prediction of pressure drops in flowing fluids, calculation of heat transfer coefficients, as well as numerous types of vapor-liquid equilibrium calculations. Programs for the prediction of properties can be used as subroutines in the design programs.

In all cases the programs are described in the text material. In addition a printout of the program and sample output are given. Storage information, that is the items which must be stored before calculations are started, is provided as well as a few program notes on specific details. The values of the physical constants used in the example are included either in the Appendix or if used in only a single program in the program notes. With this information the user may check his/her programs.

The use of the BASIC language means that the programs can easily be modified for use on the many types of personal computers which use a form of the same language.

James H. Weber

Contents

Preface	iii
1 EQUATIONS OF STATE	1
1.1 Redlich-Kwong Equation Applied to Pure Substances	1
1.2 Redlich-Kwong Equation Applied to Binary Mixtures	10
1.3 Peng-Robinson Equation Applied to Pure Substances	15
2 VOLUMETRIC RELATIONSHIPS FOR LIQUIDS	27
2.1 Saturated and Compressed Pure Liquids	27
2.2 Saturated Liquid Mixtures	32
3 VAPOR PRESSURE - TEMPERATURE CORRELATIONS	37
3.1 Lee-Kesler Correlation	37
3.2 Antoine Equation	41
3.3 Riedel, Plank, Miller Correlation	41
3.4 Frost-Kalkwarf-Thodos Correlation	44
4 LATENT HEAT OF VAPORIZATION	45
4.1 Empirical Correlations	45
4.2 Clapeyron Equation	47
5 HEAT CAPACITY EXPRESSIONS FOR GASES	51
5.1 $C_p = a + bT + cT^2 + dT^3$	51
5.2 $C_p = a + bT + c/T^2$	56
6 CALCULATION OF TOTAL CHANGES IN ENTHALPY AND ENTROPY	59
6.1 Calculation of ΔH and ΔS	60
6.2 Calculation of Final Temperature when $\Delta S = 0$	63

7 CORRELATIONS FOR PREDICTION OF VISCOSITIES	67
7.1 Viscosities of Pure Liquids and Binaries	67
7.2 Gases	71
7.3 Binary Gaseous Mixtures	78
8 CORRELATIONS FOR PREDICTION OF THERMAL CONDUCTIVITY	83
8.1 Pure Liquids and Mixtures	83
8.2 Pure Gases - Low and High Pressure	88
8.3 Binary Gaseous Mixtures	91
9 CORRELATIONS FOR PREDICTION OF DIFFUSION COEFFICIENTS	95
9.1 Diffusion Coefficients in the Liquid Phase	95
9.2 Diffusion Coefficients for Non-Polar Gaseous Mixtures	99
9.3 Diffusion Coefficients for Polar Gaseous Mixtures	101
10 SURFACE TENSION CORRELATIONS	105
10.1 Pure Liquids - Non-polar and Polar	105
10.2 Binary Mixtures	108
11 CHEMICAL EQUILIBRIUM CONSTANTS	113
11.1 Equilibrium Reaction Constants for Ideal Gas	113
12 VAPOR LIQUID EQUILIBRIUM	119
12.1 Parameters From γ^∞ Values	120
12.2 Parameters from Azeotropic Conditions	125
12.3 Calculations for Given Compositions	125
13 FLUID FRICTION AND ORIFICE CALCULATIONS	127
13.1 Pressure Drop Calculations/Incompressible Fluids	127
13.2 Pressure Drop Calculations for Compressible Fluids	130
13.3 Sizing an Orifice	134
13.4 Determination of Liquid Flowrates	136
13.5 Determination of Gas Flowrates	138
14 HEAT TRANSFER CALCULATIONS	143
14.1 Calculation of the Overall Coefficients	143
14.2 Fluids Without Phase Change	148
14.3 Condensing Vapors	150
14.4 Shell and Tube Exchangers	152
15 EQUILIBRIUM STAGE CALCULATIONS	159
15.1 Dewpoint Calculations	159
15.2 Bubblepoint Calculations	163
15.3 Flash Calculations	164
15.4 Theoretical Plates-Stripping Section	167
15.5 Theoretical Plates-Rectifying Section and Total Reflux	172
REFERENCES	177
INDEX	181

1
Equations of State

Equations of state (EOS) are used extensively in chemical engineering calculations. These equations reflect the pressure-volume-temperature behavior of fluids. And, as such, are used in sizing pipes, compressors, and for the prediction of thermodynamic properties. Equations of state for gases are dealt with here.

Programs for two equations of state, the Redlich-Kwong and the Peng-Robinson are given. The Redlich-Kwong (R-K) equation has been applied to pure components and binary mixture, while the Peng-Robinson (P-R) has been applied to pure components only. In all cases solutions for volume, pressure or temperature are given. Subroutines for the calculation of thermodynamic properties, virial coefficients, fugacities, and liquid roots are included in a number of instances.

1.1 REDLICH-KWONG EQUATION APPLIED TO PURE SUBSTANCES

The Redlich-Kwong (R-K) equations of state (1949)

$$P = RT/(V - b) - a/(T^{.5} V(V + b)) \qquad (1.1)$$

is often used in its original or in a modified form. In the original expression the two parameters, determined from conditions at the critical point, are:

$$a = 0.4278 \, R^2 \, T_c^{2.5}/P_c \qquad (1.2)$$

$$b = 0.0867 \, RT_c/P_c \qquad (1.3)$$

Frequently, it is more convenient to express the constants as functions of temperature:

$$A^2 = a/(R^2 \, T^{2.5}) \qquad (1.4)$$

$$B = b/(RT) \qquad (1.5)$$

The R-K EOS may be solved for one of the variables when the other two are specified. Thus programs are given for the determination of V when P and T are known, for P when V and T are known, and for T when V and P are given. But, only the determination of P permits a direct solution; i.e., the equation is explicit in P, but implicit in V and T.

1.1.1 Solution for Volume

Program 1 is for the solution of V, given P and T. To accomplish this the R-K is re-written in the form:

$$Z^3 - Z^2 - (BP + B^2 P^2 - AP)Z - (ABP^2) = 0 \qquad (1.6)$$

where Z is the compressibility factor and is equal to PV/RT. Equation 1.6 is convenient for the use of Newton's method. The first value of Z is set at unity in the program and then modified through the expression:

$$Z1 = Z - (f(Z)/f'(Z)) \qquad (1.7)$$

Equations of State

where f(Z) is the left hand number of equation 1.6 and f'(Z) is the derivative of that quantity. These are steps 35-50 in the program. The absolute value of f(Z1) is tested against an arbitrarily selected limit--.00001 in the program. If the limit is not met, Z1 is substituted for Z and another iteration is made. After the limit is satisfied, the volume is calculated and Z and V (liters/g mole) are printed.

Program 1 is listed in Table 1.1 and a Sample output in Table 1.2. The input data are P in atm and T in °C. Constants which must be stored prior to the use of the program as well as some explanatory notes are given in Table 1.2. A flow diagram is included as Figure 1.1.

Program 1 includes a number of subroutines. Their use is optional. These are for the calculation of:

1. Isothermal enthalpy changes
2. Isothermal entropy changes
3. Fugacity coefficient and fugacity
4. 2nd, 3rd, and 4th virial coefficients
5. Liquid volume (if any)

The isothermal enthalpy change is:

$$(H-H^*)_T = (3a/(2bT^{.5}))\ln(V/(V+b)) - RT + PV \tag{1.8}$$

Equation 1.8 was obtained from the expression:

$$(H-H^*)_T = -\int_{V^o}^{V} (P-T(\partial P/\partial T)_V)\,dV - d(PV) \tag{1.9}$$

and the R-K equation. Where H* is the enthalpy of the substance in the ideal gas state (V^o) and H the enthalpy in the real gas state at a specified volume, V. Similarily:

$$(S-S^*)_T = R(\ln((V-b)/V) - (a/(2bT^{1.5})))$$
$$\ln((V+b)/B) + R\ln(V/(RT)) \tag{1.10}$$

Table 1.1
Program 1
Solution of the Redlich-Kwong EOS When Volume Is the
Unknown Variable and Subroutines for Calculating
$(H-H^*)_T$, $(S-S^*)_T$ f/P, Fugacity, Liquid Volume, and
2nd, 3rd and 4th Virial Coefficients

```
1:"A"INPUT P,T
4:PRINT "P=";P
  ;"-ATM-";"T=
  ";T;"-C"
5:Z=1:T=T+A(29
  )
10:A(30)=T/A(27
   )
15:A=(.4278/(A(
   30)^2.5*A(28
   ))
20:B=(.0867/(A(
   30)*A(28))
25:C=(B*P+B^2*P
   ^2-A*P)
30:D=(A*P^2*B)
35:G=Z^3-Z^2-C*
   Z-D
40:F=3*Z^2-2*Z-
   C
45:A(31)=Z-(G/F
   )
50:I=A(31)^3-A(
   31)^2-C*A(31
   )-D
55:IF ABS (I)<.
   00001GOTO 10
   0
60:Z=A(31)
61:PAUSE USING
   ;Z
65:GOTO 35
100:Z=A(31):V=(Z
    *R*T)/P
105:PRINT "Z=";Z
    ;"V=";V;"--L
    ITERS/GMOL"
106:GOSUB 150
107:GOSUB 200
108:GOSUB 230
110:END

150:C=A*R^2*T^2.
  5:D=B*R*T
155:Q=V/(V+D)
160:H=(((3*C)/(2
    *D*T^.5))*LN
    (Q)-R*T+P*V)
    *24.2179
165:S=R*(LN ((V-
    D)/V))-(C/(2
    *D*T^1.5))*
    LN (1/Q)+R*
    LN (V/(R*T))
166:S=S*24.2179
170:M=(Z-B*P):N=
    (1+(B*P)/Z)
175:F=(Z-1-LN (M
    )-((A/B)*LN
    (N)))
180:F=EXP (F):J=
    F*P
185:PRINT "H=";H
    ;"--CAL/GMOL
    E-";"S=";S;"
    --CAL/GMOL-K
    --";"F/P=";F
186:PRINT "FUG="
    ;J;"--ATM"
190:RETURN
200:A(32)=D-(C/(
    R*T^1.5))
205:A(33)=D^2+((
    C*D)/(R*T^1.
    5))
210:A(34)=D^3-((
    C*D^2)/(R*T^
    1.5))
215:PRINT "2ND="
    ;A(32);"--L/
    GMOLE"
216:PRINT "3RD="
    ;A(33);"--(L
    /GMOL)^2"
217:PRINT "4TH="
    ;A(34);"--(L
    /GMOL)^3"
220:RETURN

230:K=(ABS (Z-1)
    ^2-(4*(Z*(Z-
    1)+(A*P-B^2*
    P^2-B*P))))
235:IF K<0GOTO 2
    70
240:A(35)=((1-Z)
    -K^.5)/2
245:Z=A(35)
250:V=(Z*R*T)/P
252:PRINT "ZL=";
    Z;"VL=";V
255:GOSUB 150
260:GOTO 275
270:PRINT "OTHER
    -ROOTS-ARE-I
    MAGINERY"
275:END
```

Table 1.2
Program 1
Sample Output, Input Data Storage
Information, and Program Notes

```
P=1.-ATM-T=-161.
45-C
Z=0.966441289V=8
.859292494--LITE
RS/GMOL
H=-19.77571885--
CAL/GMOLE-S=-1.1
13266153E-01--CA
L/GMOL-K--F/P=9.
674805791E-01
FUG=9.674805791E
-01--ATM
2ND=-2.985935638
E-01--L/GMOLE
3RD=1.070291838E
-02--(L/GMOL)^2
4TH=-2.6638867E-
04--(L/GMOL)^3
ZL=4.460529805E-
03VL=4.088933148
E-02
H=-2229.25165--C
AL/GMOLE-S=-19.3
5435868--CAL/GMO
L-K--F/P=7.38901
4211E-01
FUG=7.389014211E
-01--ATM
```

Output of Program 1

Input Data:

P = 1 atm
t = -161.45 $^{\circ}$C
Note: Normal boiling point of methane

Storage Information

R = gas law constant (0.0826 liter/atm/g mol - $^{\circ}$K)
A(27) = T_c, $^{\circ}$K
A(28) = P_c, atm
A(29) = 273.16 $^{\circ}$K (0°C)

Program Notes

1. Step 61 is "Pause Using; Z". The TRS-80 on this command displays briefly the current value of Z. This permits the observation of the course of the calculations.
2. In the program f(Z) = G; f'(Z) = F; Z1 = A(31); f(Z1) = I; and f(fugacity) - FUG.
3. Input data are P in atm and T in $^{\circ}$C, although any consistent set of units may be used.
4. At the input conditions, the R-K EOS does give vapor and liquid roots. But, it does not predict exactly the point on the vapor pressure-temperature relationship, as the fugacities of the two phases are not equal.
5. Physical data for methane were used in the example.

In equation 1.10, the reference pressure was set at 1 atm. The fugacity coefficient was calculated from:

$$\ln(f/P) = (Z - 1 - \ln(1 + (BP/Z))) - ((A/B)\ln(Z - BP)) \quad (1.11)$$

The R-K equation can be re-written in virial form:

$$Z = 1 + \beta(T)/V + \gamma(T)/V^2 + \delta(T)/V^3 \quad (1.12)$$

where $\beta(T)$, $\gamma(T)$, $\delta(T)$ are the 2nd, 3rd and 4th virial coefficients, are functions of temperature only, and are:

$$\beta(T) = b - (a/RT^{1.5}) \quad (1.13)$$

$$\gamma(T) = b^2 + (ab/RT^{1.5}) \quad (1.14)$$

$$\delta(T) = b^3 - (ab^2/RT^{1.5}) \quad (1.15)$$

Lastly, the R-K relationship is a cubic equation with respect to volume, as illustrated by equation 1.6. After the initial root of equation 1.6 is determined, a quadratic equation can be obtained. If the quantity designated as K:

$$K = \text{abs}(Z - 1)^2 - (4(Z(Z - 1) + (AP - B^2P^2 - BP))) \quad (1.16)$$

is negative (step 230) the other roots are imaginary. The user is so informed and calculations cease. If K is positive, the other two roots are real. The liquid compressibility factor is the smaller of these two roots and is

$$Z_L = ((1 - Z) - K^{.5})/2 \quad (1.17)$$

This is step 240 in the program. With Z_L, V_L the isothermal enthalpy (equation 1.8) and entropy (equation 1.9) changes and the fugacity

Equations of State

coefficient and fugacity (equation 1.11) for the liquid phase can be calculated using the subroutines given previously.

Two additional points are: first, in the equation 1.16 (step 230), the quantity (Z-1) is prefixed with abs (meaning absolute value). This is necessary, since the TRS-80 yields a negative number when a negative number is squared. Hence, the absolute value of (Z-1) is used. The second point is in determining the liquid phase compressibility, a small difference between two relatively small numbers is calculated. Due to differing rounding off techniques in various computers, the small difference,--in this case Z_L--may differ rather significantly despite the fact that the input data are identical.

1.1.2 Solution for Pressure or Temperature

Program 2 offers solutions of the R-K EOS when either P or T is the unknown variable. The R-K is explicit in P, hence a direct solution is possible. The program defined as "A" in Table 1.3 gives this solution. Sample output is in Table 1.4. The storage is the same as for Program 1.

Also in Program 2 and defined as "B" in the program for the solution of T. A trial and error solution is required. The initial estimate of T is obtained from the ideal gas law,

$$T = PV/R \qquad (1.18)$$

The Newton technique is then used. Using the initial estimate of T, the pressure, P1, is calculated (step 70) and the value of the derivative, dP/dT (step 75). The next estimate of the temperature, T1 (step 80) is calculated by:

$$T1 = T - (P1 - P)/(dP/dT) \qquad (1.19)$$

T1 is then used to calculate a second value for the pressure, P2

Table 1.3
Program 2
Solution of Redlich-Kwong EOS When
Either Pressure or Temperature Is the Unknown Variable

"A"

```
1:"A"INPUT V,T
4:PRINT "V=";V
 ;"T=";T
5:T=T+A(29):A(
 30)=T/A(27)
10:A=.4278/(A(3
 0)^2.5*A(28)
 )
15:B=.0867/(A(3
 0)*A(28))
20:C=A*R^2*T^2.
 5:D=B*R*T
25:P=((R*T)/(V-
 D))-(C/(T^.5
 *V*(V+D)))
30:Z=(P*V)/(R*T
 )
35:PRINT "Z=";Z
 ;"P=";P;"--A
 TM"
45:END
```

"B"

```
50:"B"INPUT V,P
51:PRINT "V=";V
 ;"P=";P
55:T=(P*V)/R
60:C=(.4278*R^2
 *A(27)^2.5)/
 A(28)
65:D=(.0867*R*A
 (27))/A(28)
70:O=((R*T)/(V-
 D))-(C/(T^.5
 *V*(V+D)))
75:N=(R/(V-D))+
 (C/(2*T^1.5*
 V*(V+D)))
80:U=T-(O-P)/N
85:Q=((R*U)/(V-
 D))-(C/(U^.5
 *V*(V+D)))
90:IF ABS (Q-O)
 <.0005*PGOTO
 125
95:T=U
96:PAUSE USING
 ;T
100:GOTO 70
125:P=Q:T=U:Z=(P
 *V)/(R*T)
126:T=T-A(29)
130:PRINT "Z=";Z
 ;"P=";P;"T="
 ;T;"--CENTI"
131:T=T+A(29)
132:A=C/(R^2*T^2
 .5):B=D/(R*T
 )
140:END
```

Equations of State

Table 1.4
Program 2
Sample Output, Input Data Storage Information,
and Program Notes

"A"	"B"
	V=25.46P=1.
	Z=9.984063766E-0
V=25.46T=37.73	1P=9.999999999E-
Z=9.984081569E-0	01T=37.5960128--
1P=1.000432949--	CENTI
ATM	V=0.5891P=40.83
V=0.5891T=37.73	Z=9.398638178E-0
Z=9.393303281E-0	1P=40.82999995T=
1P=40.67874889--	38.7088212--CENT
ATM	I
V=0.1279T=37.73	V=0.1279P=170.11
Z=8.542711634E-0	Z=8.536374953E-0
1P=170.3975977--	1P=170.11T=37.43
ATM	56676--CENTI
Program for P = ?	Program for T = ?

Input Data:

"A" V = (three values) t = 37.73°C
"B" V = (three values) P = (three values)

Storage Information: (same for both solutions)

```
R     = gas constant (.08206 liter-atm/g mole-°K)
A(27) = T_c, °K
A(28) = P_c, atm
A(29) = 273.16 °K (0°C)
```

Program notes:
1. Step 96 "Pause Using; T" allows observation of the course of the calculations.
2. In program "B" P1 = 0; dP/dT = N; T1 = U, and P2 = Q.
3. Physical data for methane were used in the examples and are given in the Appendix.

(step 85). If the absolute difference between the two calculated pressure values (P2 - P1) is less than an arbitrary limit (0.0005 P in this case) a satisfactory solution has been obtained. If not, the new value of the temperature, T1, is set equal to T and the calculations repeated. A satisfactory solution is usually reached within four tries.

When the results are acceptable, Z is calculated and printed as are the calculated P and T. The program is given in Table 1.3 and a sample output in Table 1.4. A flow chart is included in Figure 1.2. The subroutines given with Program 1 may be used with these programs.

1.2 REDLICH-KWONG EQUATION APPLIED TO BINARY MIXTURES

The R-K equation may be applied to mixtures as well as pure components. While the form of the equation remains the same and the solutions are identical, the constants are a function of an additional variable; namely composition, and the critical temperature and pressure must be redefined. These differences must be taken into account.

$$T_{cm} = ((\Sigma\ y_i (T_{ci}^{5/2}/P_{ci})^{1/2})^2 / \Sigma\ y_i (T_{ci}/P_{ci}))^{2/3} \tag{1.20}$$

$$P_{cm} = T_{cm} / \Sigma\ y_i (T_{ci}/P_{ci}) \tag{1.21}$$

These equations follow from the combination of the constants of the pure components to determine mixture constants. For a binary mixture the recommended combinations are:

$$a_m = y_1^2 a_{11} + y_2^2 a_{22} + 2 y_1 y_2 a_{12} \tag{1.22}$$

Equations of State

$$a_{12} = (a_1 a_2)^{1/2} \tag{1.23}$$

$$b_m = y_1 b_1 + y_2 b_2 \tag{1.24}$$

$$A_m = y_1 A_1 + y_2 A_2 \tag{1.25}$$

$$B_m = y_1 B_1 + y_2 B_2 \tag{1.26}$$

After the mixture constants are obtained. The method of solution follows the same course as described in Program 1. The input data requires a value of y, the mole fraction of component one, in addition to the variables of T, V, or P.

1.2.1 Solution for Volume

Program 3, Table 1.5, is the solution of the R-K EOS for a binary mixture for volume when pressure and temperature are given. The method employed is similar to that used in Program 1. The Newton convergence technique includes steps 45-60. Also in the program are subroutines for the calculation of $(H-H^*)_T$ and $(S-S^*)_T$, the isothermal changes in enthalpy and entropy for the mixture of composition, y. The entropy of mixing term is not included in program. Sample output, storage information and program notes are in Table 1.6.

1.2.2 Solution for Pressure or Temperature

Program 4, Table 1.7, gives the solution of the R-K equation when either P or T is the unknown quantity. These solutions are similar to those used for a pure component. The differences are that the constants and critical properties are functions of composition as shown in Equation 1.20 to 1.26.

Since there is a great deal in common in the calculations when either P or T is the unknown variable a single program is given for

Table 1.5
Program 3
Solution of the Redligh-Kwong EOS as Applied to a
Binary Mixture When V Is the Unknown Variable
and Subroutines for Calculation $(H-H^*)_T$ and $(S-S^*)_T$

```
1:"A"INPUT P,T           27:A(48)=(.4278         150:C=A(42)*Y^2+
  ,Y                        *A(40)^2.5)/            A(43)*X^2+2*
2:PRINT "P=";P              (A(41)*T^2.5            X*Y*(A(42)*A
  ;"T=";T;"Y1=              )                       (43))^.5
  ";Y                    30:A(51)=(.0867         155:D=Y*A(44)+X*
5:T=T+A(29)                 *A(40))/(A(4            A(45):Q=V/(V
6:X=1-Y                     1)*T)                   +D)
10:A(40)=((Y*(A          33:A=A(48):B=A(         160:H=(((3*C)/(2
  (27)^2.5/A(2              51)                     *D*T^.5))*LN
  8))^.5)+(X*(          35:C=(B*P+B^2*P             (Q)-R*T+P*V)
  A(32)^2.5/A(              ^2-A*P)                 *24.2179
  33))^.5)^2           40:D=(A*P^2*B):         165:S=(R*(LN ((V
12:A(40)=(A(40)             Z=1                     -D)/V))-(C/(
  /((Y*(A(27)/         45:G=Z^3-Z^2-C*             2*D*T^1.5))*
  A(28)))+(X*(             Z-D                     LN (1/Q)+R*
  A(32)/A(33))         50:F=3*Z^2-2*Z-             LN (V/(R*T))
  ))^(2/3)                  C                       )*24.2179
15:A(41)=A(40)/         55:A(31)=Z-(G/F         170:PRINT "H=";H
  ((Y*(A(27)/A              )                       ;"--CAL/GMOL
  (28)))+(X*(A         60:I=A(31)^3-A(             E";"S=";S;"-
  (32)/A(33)))             31)^2-C*A(31             -CAL/GMOLE-K
  )                        )-D                      "
20:A(42)=(.4278         65:IF ABS (I)<.         175:RETURN
  *R^2*A(27)^2              00001GOTO 10
  .5)/A(28)                 0
21:A(43)=(.4278         70:Z=A(31)
  *R^2*A(32)^2         71:PAUSE USING
  .5)/A(33)                 ;Z
24:A(44)=(.0867         75:GOTO 45
  *R*A(27))/A(        100:Z=A(31):V=(Z
  28):A(45)=(.              *R*T)/P
  0867*R*A(32)        105:PRINT "Z=";Z
  )/A(33)                 ;"V=";V;"--L
                          ITERS/GMOLE"
                      107:GOSUB 150
                      110:END
```

the two types of problems. The first instruction in the program informs the user to set the unknown variable at zero, when P, T, V and y are entered. So, in the calculations the common steps are carried out and then using the fact that either P or T is unknown (therefore zero) the calculations are given the proper direction - step 28. Subroutines for the calculation of $(H-H^*)_T$ and $(S-S^*)_T$ are included. Sample output, input and other information are given in Table 1.8.

Equations of State   13

                         Table 1.6
                         Program 3
                    Output-Input Data Storage
                    Information and Program Notes

```
P=13.609T=38.61Y
1=0.477
Z=9.466001947E-0
1V=1.779533676--
LITERS/GMOLE
H=-103.6340702--
CAL/GMOLES=-5.41
5545128CAL/GMOLE
-K
```

Input Data:

  P = 13.609 atm
  T = 38.61°C
  y_1 = 0.477 (mole fraction methane in a methane-ethane mixture)

Storage Information:

  R   = gas law constant (0.08206 liter/atm/g mole °K)
  A(27) = T_{c1}, °K
  A(28) = P_{c1}, atm
  A(29) = 273.16°K (0°C)
  A(32) = T_{c2}, °K
  A(33) = P_{c2}, atm

Program Notes:

  1. In general the terminology is the same as used for the analogous program for a pure component. In addition y_2 = x, T_{CM} = A(40), P_{CM} = A(41), A_M = A(48), B_M = A(51), a_M = C, and b_M = D.
  2. Step 71 in "Pause Using Z". This allows the observation of the course of the calculations.
  3. Data on the methane-ethane system were used in the example.

Table 1.7
Program 4
Solution of Redlich-Kwong EOS as Applied to a
Binary Mixture Where Either P or T Is the Unknown Variable
and Subroutines for Calculating $(H-H^*)_T$ and $(S-S^*)_T$

```
1:"A"PRINT "SE
  T-UNKNOWN-AS
  -0"
2:INPUT P,T,V,
  Y
5:PRINT "P=";P
  ;"T=";T;"V="
  ;V;"Y1=";Y
6:X=1-Y
10:A(40)=((Y*(A
   (27)^2.5/A(2
   8))^.5)+(X*(
   A(32)^2.5/A(
   33))^.5))^2
12:A(40)=(A(40)
   /((Y*(A(27)/
   A(28)))+(X*(
   A(32)/A(33))
   )))^(2/3)
15:A(41)=A(40)/
   ((Y*(A(27)/A
   (28)))+(X*(A
   (32)/A(33)))
   )
20:A(42)=(.4278
   *R^2*A(27)^2
   .5)/A(28)
21:A(43)=(.4278
   *R^2*A(32)^2
   .5)/A(33)
25:C=A(42)*Y^2+
   A(43)*X^2+2*
   X*Y*(A(42)*A
   (43))^.5
27:D=(.0867*R*A
   (40))/A(41)
28:IF T=0GOTO 6
   0

30:T=T+A(29)
33:P=((R*T)/(V-
   D))-(C/(T^.5
   *V*(V+D)))
35:Z=(P*V)/(R*T
   )
40:PRINT "Z=";Z
   ;"P=";P
45:GOSUB 155
50:END
60:T=(P*V)/R
61:PAUSE USING
   ;T
70:O=((R*T)/(V-
   D))-(C/(T^.5
   *V*(V+D)))
75:N=(R/(V-D))+
   (C/(2*T^1.5*
   V*(V+D)))
80:U=T-(O-P)/N
85:Q=((R*U)/(V-
   D))-(C/(U^.5
   *V*(V+D)))
90:IF ABS (Q-O)
   <.0005*PGOTO
   125
95:T=U
96:PAUSE USING
   ;T
100:GOTO 70
125:P=Q:T=U:Z=(P
   *V)/(R*T)
126:T=T-A(29)
130:PRINT "Z=";Z
   ;"P=";P;"T="
   ;T;"--CENTI"
131:T=T+A(29)
132:GOSUB 155
135:END

155:Q=V/(V+D)
160:H=(((3*C)/(2
   *D*T^.5))*LN
   (Q)-R*T+P*V)
   *24.2179
165:S=(R*(LN ((V
   -D)/V))-(C/(
   2*D*T^1.5))*
   LN (1/Q)+R*
   LN (V/(R*T))
   )*24.2179
170:PRINT "H=";H
   ;"--CAL/GMOL
   E";"S=";S;"-
   -CAL/GMOLE-K
   "
175:RETURN
```

Equations of State

Table 1.8
Program 4
Sample Output, Input Data, Storage
Information and Program Notes

```
SET-UNKNOWN-AS-0            SET-UNKNOWN-AS-0
P=0.T=38.61V=1.7            P=68.045T=0.V=0.
8Y1=0.477                   2708Y1=0.477
Z=9.466136439E-0            Z=7.201276542E-0
1P=13.60562802              1P=68.04499999T=
H=-103.607448--C            38.6599351--CENT
AL/GMOLES=-5.414            I
993561--CAL/GMOL            H=-611.9480984--
E-K                         CAL/GMOLES=-9.80
                            7070255--CAL/GMO
                            LE-K

        P = ?                      T = ?
```

Input Data:

P = ?   P = 0; t = 38.61°C; V = 1.78 liters/g mole; y_1 = 0.477

T = ?   P = 68.045 atm; t = 0; V = 1.78 liters/g mole; y_1 = 0.477

Storage Information same as Program 3.

Program Notes:

1. Mixture constant are the same as used in Program 3.
2. Method of calculation and symbols are identical to those used in Program 2.
3. Data on the methane-ethane system were used in the example.

1.3 PENG-ROBINSON EQUATION APPLIED TO PURE SUBSTANCES

The Peng-Robinson EOS (1976):

$$P = RT/(V - b) - a(T)/(V(V + b) + b(V - b)) \qquad (1.27)$$

is similar to the Redlich-Kwong in that it is explicit in P, but implicit in V and T and has two constants. However, the P-R represents a step forward in that the "a" term is a function of temperature and the acentric factor and the term b(V-b) in the denominator of the second term on the RHS are added. Hence, overall a better fit of experimental data is obtained.

The constants for the Peng-Robinson EOS are evaluated as follows:

$$a(T) = a(T_c)\, \alpha(T_r, \omega) \qquad (1.28)$$

$$b(T) = b(T_c) \qquad (1.29)$$

$$a(T_c) = 0.45724\, (R^2 T_c^2 / P_c) \qquad (1.30)$$

$$b(T_c) = 0.07780\, (RT_c / P_c) \qquad (1.31)$$

$$\alpha^{1/2} = 1 + K(1 - T_r)^{1/2} \qquad (1.32)$$

$$K = 0.37464 + 1.54226\, \omega - 0.26992\, \omega^2 \qquad (1.33)$$

1.3.1 Solution for Volume

To solve the P-R EOS for volume, temperature and pressure known, the form:

$$Z^3 - (1 - B)Z^2 + (A - 3B^2 - 2B)Z - (AB - B^2 - B^3) = 0 \qquad (1.34)$$

where

$$A = aP/R^2 T^2 \qquad (1.35)$$

$$B = bP/RT \qquad (1.36)$$

was used. The technique was the same as used for the R-K and a flow diagram would be similar to Figure 1.1. As before the initial value of the compressibility factor was set at unity in the program. The value was modified by the Newton technique, Equation 1.7, and when the absolute value (LHS of Equation 1.34) of f(Z1) was within an arbitrarily defined limit, 0.00001 in this case, a satisfactory

Equations of State

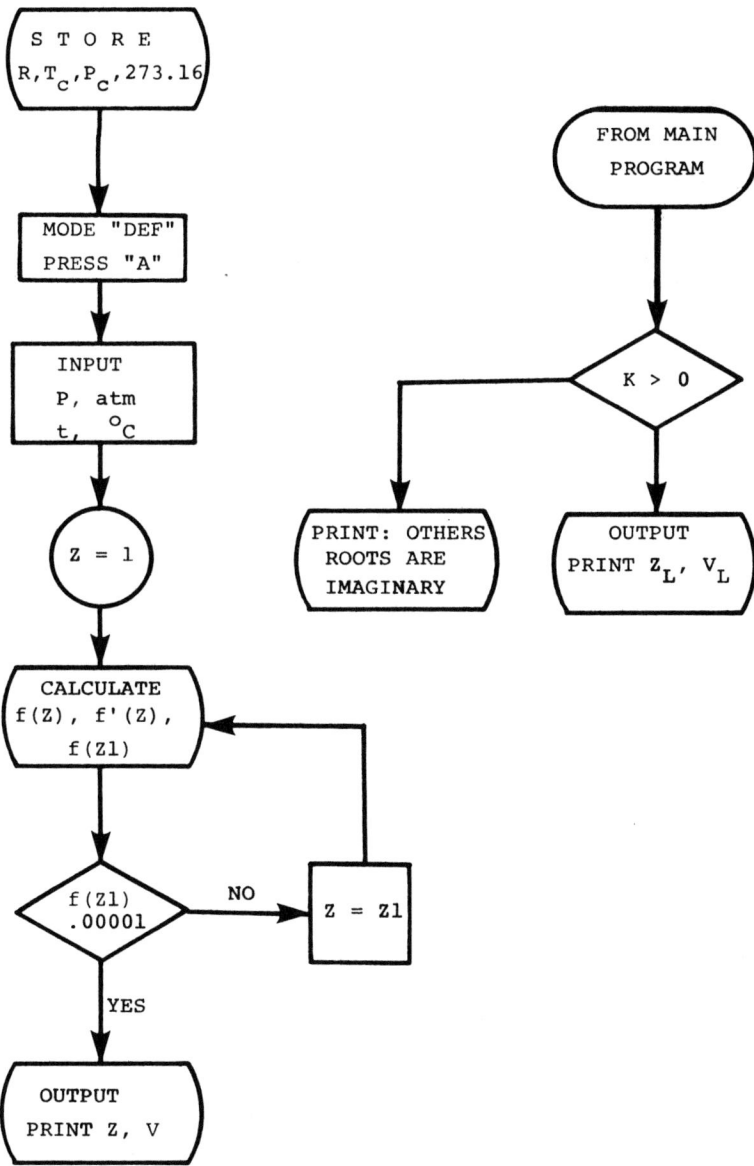

Figure 1.1

solution has been obtained. These are step 30-55 in the program. V is then calculated and Z and V printed.

Program 5 is listed in Table 1.9 and a sample output, input data, storage information and program notes are given in Table 1.10.

Program 5 also includes a number of subroutines which may be used. These are to calculate:

1. Isothermal enthalpy changes
2. Isothermal entropy changes
3. Fugacity coefficients and fugacities
4. 2nd, 3rd and 4th virial coefficients
5. Liquid root (if any).

The relationships used are:

$$(H - H^*)_T = RT(Z - 1) \frac{T(\partial a/\partial T) - a}{2^{1.5} b} \ln \left(\frac{Z + 2.414B}{Z - 0.414B}\right) \tag{1.37}$$

$$(S - S^*)_T = R \ln\left(\frac{V - b}{RT}\right) + \frac{\partial a/\partial T}{2^{1.5} b} \ln \left(\frac{Z + 2.414B}{Z - 0.414B}\right) \tag{1.38}$$

$$\ln \frac{f}{p} = Z - 1 - \ln(Z - B) - \frac{A}{2^{1.5} B} \ln \left(\frac{Z + 2.414B}{Z - 0.414B}\right) \tag{1.39}$$

For the virial coefficients:

$$\beta(T) = b - a(T)/RT \tag{1.40}$$

$$\gamma(T) = b^2 + 2a(T)b/RT \tag{1.41}$$

$$\delta(T) = b^3 - 5a(T)b^2/RT \tag{1.42}$$

Since the P-R is a cubic equation with respect to volume, it will predict a liquid root. If the second and third roots are real the quantity Q, step 160, as defined:

$$Q = abs(Z - (1 - B))^2 - (4(Z(Z - (1 - B)) + Y)) \tag{1.43}$$

Table 1.9
Program 5
Solution of the Peng-Robinson EOS When the
Volume Is the Unknown Variable and
Subroutines for $(H-H^*)_T$, $(S-S^*)_T$, 2nd, 3rd and
4th Virial Coefficients, Fugacity and Liquid Root

```
1:"A"INPUT P,T
2:PRINT "P=";P
  ;"T=";T
5:T=T+A(29)
10:C=(.45724*R^
   2*A(27)^2)/A
   (28):D=(.077
   8*R*A(27))/A
   (28)
15:K=.37461+(1.
   54226*W)-(.2
   6992*W^2)
17:A(30)=T/A(27
   )
20:E=C*((1+K*(1
   -A(30)^.5))^
   2
25:A=(E*P)/(R^2
   *T^2):B=(D*P
   )/(R*T)
30:X=(A*B-B^2-B
   ^3):Y=(A-3*B
   ^2-2*B):Z=1
35:G=Z^3-((1-B)
   *Z^2)+Y*Z-X
40:I=3*Z^2-(2*(
   1-B)*Z+Y
45:A(31)=Z-(G/I
   )
50:J=A(31)^3-((
   1-B)*A(31)^2
   )+Y*A(31)-X
55:IF ABS (J)<.
   00001GOTO 80
60:Z=A(31)
61:PAUSE USING
   ;Z
65:GOTO 35
80:Z=A(31):V=(Z
   *R*T)/P
85:PRINT "Z=";Z
   ;"V=";V;"--L
   ITERS/GMOLE"
87:GOSUB 110
89:GOSUB 150
91:GOSUB 160
93:GOSUB 110
95:END
110:L=Z-B:M=(Z+(
    2.414*B))/(Z
    -(.414*B))
115:F=Z-1-LN (L)
    -((A/(2^1.5*
    B))*LN (M)
120:F=EXP (F):N=
    F*P
125:PRINT "F/P="
    ;F;"FUGACITY
    =";N
130:A(40)=R*T*(Z
    -1):A(41)=(-
    K*C*A(30)^.5
    )
132:A(42)=(K^2*C
    *A(30)^.5):A
    (43)=(K^2*C*
    A(30))
135:H=(A(40)+(((
    A(41)-A(42)+
    A(43))-E)/(2
    ^1.5*D)*LN (
    M)))*24.2179
139:A(44)=K*A(27
    )^-.5-T^-.5-
    K*T^-.5
141:S=(R*LN ((V-
    D)/(R*T))+((
    (C*K*A(27)^-
    .5)/(2^1.5*D
    ))*A(44)*LN
    (M)))*24.217
    9
145:PRINT "H=";H
    ;"--CAL/GMOL
    E";"S=";S;"-
    -CAL/GMOLE-K
    "
146:RETURN
150:A(35)=D-(E/(
    R*T)):A(36)=
    D^2+((2*E*D)
    /(R*T))
155:A(37)=D^3-((
    5*E*D^2)/(R*
    T))
157:PRINT "2ND="
    ;A(35);"3RD=
    ";A(36);"4TH
    =";A(37);"--
    VIRIAL-COEFT
    S"
158:PRINT "LITER
    S/GMOLE"
159:RETURN
160:Q=ABS ((Z-(1
    -B)))^2-(4*(
    Z*(Z-(1-B))+
    Y))
165:IF Q<0GOTO 2
    00
170:Z=(-(Z-(1-B)
    )-Q^.5)/2
175:V=(Z*R*T)/P
180:PRINT "ZL=";
    Z;"VL=";V
185:RETURN
200:PRINT "OTHER
    -ROOTS-ARE-I
    MAGINARY"
202:END
```

Table 1.10
Program 5
Sample Output, Input Data, Storage
Information and Program Notes

```
P=1.T=-161.45
Z=9.672611271E-0
1V=8.866807886--
LITERS/GMOLE
F/P=0.968217484F
UGACITY=0.968217
484
H=-17.39334243--
CAL/GMOLES=-9.15
132848E-02--CAL/
GMOLE-K
2ND=-2.922890777
E-013RD=1.777951
055E-024TH=-1.12
1780067E-03--VIR
IAL-COEFTS
LITERS/GMOLE
ZL=3.720329487E-
03VL=3.410397245
E-02
F/P=9.950870771E
-01FUGACITY=9.95
0870771E-01
H=-1953.327652--
CAL/GMOLES=-17.4
7591332--CAL/GMO
LE-K
```

Input Data:

P = 1 atm
t = -161.45°C
normal boiling point of methane

Storage Information:

R   = gas law constant (0.0826) liter-atm/g mole °K)
W   = acentic factor
A(27) = T_c, °K
A(28) = P_c, atm
A(29) = 273.16°K (0°C)

Program Notes:

1. Solution, in general, the same as Program 1.
2. Physical data for methane used in examples.

Equations of State

Table 1.11
Program 6
Solution of Peng-Robinson EOS when the
Pressure is the Unknown Variable

```
1:"A"INPUT T,V
5:PRINT "T=";T
;"V=";V
10:C=(.45724*R^
2*A(27)^2)/A
(28):D=(.077
8*R*A(27))/A
(28)
15:K=.37461+(1.
54226*W)-(.2
6992*W^2)
17:U=(V*(V+D))+
(D*(V-D))
25:T=T+A(29):A(
30)=T/A(27)
30:E=C*((1+K*(1
-A(30)^.5))^
2
40:P=((R*T)/(V-
D))-(E/U)
45:A=(E*P)/(R^2
*T^2):B=(D*P
)/(R*T)
50:X=(A-3*B^2-2
*B)
51:Z=(P*V)/(R*T
):Y=(A-3*B^2
-2*B)
52:PRINT "Z=";Z
;"P=";P;"--A
TM"
54:GOSUB 110
55:GOSUB 150
56:GOSUB 160
57:GOSUB 110
60:END
```

where

$$Y = A - 3B^2 - 2B \qquad (1.44)$$

must be greater than zero. Note that it is necessary to use the absolute value of the first term in equation 1.43. If real roots exist the smaller of the two is the liquid root and is obtained (step 170) by:

Table 1.12
Program 6
Sample Output, Input Data, Storage Information

```
T=-161.45V=8.869
Z=9.672691752E-0
1P=0.999760119--
ATM
F/P=9.682250847E
-01FUGACITY=9.67
9928259E-01
H=-17.38928324--
CAL/GMOLES=-9.10
1577094E-02--CAL
/GMOLE-K
2ND=-2.922890777
E-013RD=1.777951
055E-024TH=-1.12
1780067E-03--VIR
IAL-COEFTS
LITERS/GMOLE
ZL=3.676496178E-
03VL=3.371024235
E-02
F/P=9.944067478E
-01FUGACITY=9.94
1682085E-01
H=-1969.106054--
CAL/GMOLES=-17.6
1532168--CAL/GMO
LE-K
```

Input Data:

t = -161.45°C
V = 8.869 liters/g mole
Normal boiling point of methane

Storage information same as Program 5

Program Notes:

1. Subroutines used with Program 5 may be used with this program.

Equations of State

Table 1.13
Program 7
Solution of Peng-Robinson EOS when the Temperature is the Unknown Variable

```
1:"A"INPUT P,V
5:PRINT "P=";P
  ;"V=";V
10:C=(.45724*R^
   2*A(27)^2)/A
   (28):D=(.077
   8*R*A(27))/A
   (28)
15:K=.37461+(1.
   54226*W)-(.2
   6992*W^2)
17:U=(V*(V+D))+
   (D*(V-D))
20:T=(P*V)/R
25:A(30)=T/A(27
   )
30:E=C*((1+K*(1
   -A(30)^.5))^
   2
40:Q=((R*T)/(V-
   D))-(E/U)
45:A(45)=((-K/(
   T^.5*A(27)^.
   5))-(K^2/(T^
   .5*A(27)^.5)
   )+(K^2/A(27)
   ))
50:G=(R/(V-D))-
   ((E/U)*A(45)
   )
51:A(46)=T-(Q-P
   )/G
52:T=A(46)
53:A(30)=T/A(27
   )
54:PAUSE USING
   ;T
55:E=C*((1+K*(1
   -A(30)^.5))^
   2
56:A(45)=((-K/(
   T^.5*A(27)^.
   5))-(K^2/(T^
   .5*A(27)^.5)
   )+(K^2/A(27)
   ))
57:Q=((R*T)/(V-
   D))-(E/U)
60:IF ABS (Q-P)
   <.0005*PGOTO
   75
65:GOTO 50
75:P=Q:Z=(P*V)/
   (R*T)
80:A=(E*P)/(R^2
   *T^2):B=(D*P
   )/(R*T)
81:X=(A-3*B^2-2
   *B):A(30)=T/
   A(27)
85:T=T-A(29)
86:PRINT "Z=";Z
   ;"P=";P;"T="
   ;T
87:T=T+A(29)
90:GOSUB 110
95:END
```

$$Z_L = (-(Z-(1-B))-Q^{1/2})/2 \qquad (1.45)$$

with Z_L determined, V_L can be calculated. The quantities are printed. If Q is negative the message "other roots are imaginary" is printed and calculations cease. Also, if the liquid phase exists, $(H-H^*)_T$, $(S-S^*)_T$, f/P ratio and fugacity can be calculated for this phase using the subroutines previously described.

1.3.2 Solution for Pressure

The Peng-Robinson (EOS) is explicit in pressure; hence, this variable can be calculated directly. Program 6, given in Table 1.11,

Table 1.14
Program 7
Sample Output, Input Data, Storage Information
and Program Notes

```
P=1.V=8.869
Z=9.672755288E-0
1P=0.999929442T=
-161.4319298
F/P=9.682319839E
-01FUGACITY=9.68
1636674E-01
H=-17.38860355--
CAL/GMOLES=-9.13
3522357E-02--CAL
/GMOLE-K
```

Input Data:
    P = 1 atm
    V = 8.869 liters/g mole

Storage information same as program 5

Program Notes:
    1. Subroutines given with Program 5 may be used with program.
    2. Physical data for methane used in example.

---

offers this solution. All the subroutines used with Program 4 may be used with this program. A sample output, input data, storage information, and program notes are given in Table 1.12.

1.3.3 Solution for Temperature

Program 7, Table 1.13, provides the solution for the P-R EOS when the temperature is the unknown variable. The solution is similar to that outlined for the R-K. The initial estimate of the temperature is made through the ideal gas law, equation 1.18, and successive temperature estimates by the Newton technique, equation 1.19, until satisfactory agreement between the given and calculated pressure is obtained. In this case the limit was arbitrarily set at 0.005P. These are steps 20 through 60 in the program. When the calculated pressure value is acceptable, this value, the compressibility factor, and temperature are printed. A sample output is given in Table 1.14, along with storage information and program notes. The flow diagram for this program is similar to Figure 1.2.

Equations of State

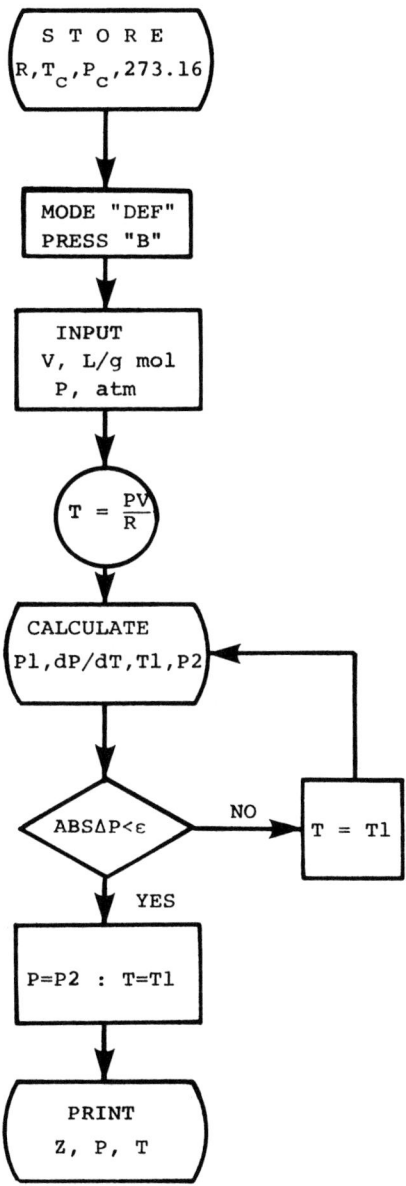

Figure 1.2

The subroutines described in Program 5 may be used with this program. However, the length of the program is such that not all subroutines may be used at one time. The capacity of the TRS-80 would be exceeded. Since the subroutines are completely independent of each other, reducing the number does not adversely effect the operation of the program. For example, if only gas phase properties are desired, the subroutines for the determination of liquid roots can be eliminated.

2
Volumetric Relationships for Liquids

Correlations to predict the volumetric behavior of liquids are less well developed than comparable expressions for gases. Results obtained by these correlations, however, are generally more accurate than those calculated by applying equations of state, developed for the gases, to the liquid phase. Also, the correlations are much less complicated and require a minimum of input data. Programs are given for the calculation of volumes of saturated and compressed pure liquids and saturated liquid mixtures.

2.1 SATURATED AND COMPRESSED PURE LIQUIDS

For pure saturated liquids programs are given for the Spencer and Danner (1972) modification of the Rackett equation (1970), the Gunn-Yamada correlation (1973), the modification of the Guggenheim relationship (1945) proposed by Lu, Ruether, Hsi, and Chiu (1973) and the Lyckman, Eckert, and Prausnitz correlation (1965). For compressed pure liquids, the relationship proposed by Chueh and Prausnitz (1967, 1969) has been programmed.

The Rackett equation as modified by Spencer and Danner is:

$$V^s = (RT_c/P_c)Z_{RA}\left[1 + (1-T_r)^{2/7}\right] \tag{2.1}$$

where Z_{RA} is an adjustable parameter and is empirically determined to get the best fit on the data. Z_{RA} varies slightly from Z_c the critical compressibility factor. The latter can be used if a value of the former is not available, but the results are not as accurate. The Gunn-Yamada relationship uses two adjustable parameters, V_{scr} and Z_{cr}, but the form of the equation is similar to the Rackett equation:

$$V^s = V_{scr} Z_{cr}^{(1-T_r)^{2/7}} \tag{2.2}$$

Values of V_{scr} are reported by the authors, but may be calculated by

$$V_{scr} = (RT_c/P_c)(0.292 - 0.0967\omega) \tag{2.3}$$

and Z_{cr} is calculated by the generalized relationship:

$$Z_{cr} = 0.29056 - 0.0877\omega \tag{2.4}$$

The correlation of Lu et al. is:

$$\rho_{rs} = 1 + \alpha(1-T_r) + \beta(1-T_r)^{1/3} \tag{2.5}$$

where

$$\alpha = 0.73098 + 0.29808\omega \tag{2.6}$$

$$\beta = 1.75238 + 0.74293\omega \tag{2.7}$$

Volumetric Relationships for Liquids

The saturated volume is obtained by:

$$V^S = V_c/\rho_{rs} \tag{2.8}$$

Lyckman, Eckert, and Prausnitz proposed the equation:

$$\frac{V^S}{V_c} = V_r^{(0)} + \omega V_r^{(1)} + \omega^2 (V_r)^{(2)} \tag{2.9}$$

where

$$V_r^{(j)} = a_j + b_j T_t + c_j T_r^2 + d_j T_r^2 + d_j T_r^3 + \frac{e_j}{T_r} + f_j \ln(1 - T_r) \tag{2.10}$$

Each of the V_r terms on the right hand side of the equation 2.9 contained six constants, hence eighteen empirical constants are used. The programs for the four correlations are included in Program 8, Table 2.1. Sample output, input data, storage information, and program notes are given in Table 2.2. The program is defined as "A". Also given in Program 8 is a program for the calculation of the volume of compressed liquids by the relationship proposed by Chueh and Prausnitz:

$$1/V = 1/V^S [1 + (9Z_c N (P - P^S)/P_c)]^{1/9} \tag{2.11}$$

where

$$N = (1.0 - 0.89\omega)[\exp(6.9457 - 76.2853 T_r + 191.306 T_r^2 - 203.547 T_r^3 + 82.7631 T_r^4)] \tag{2.12}$$

V is the liquid volume at pressure, P, which is greater than the

Table 2.1
Program 8
Determination of Saturated and Compressed Volumes
of Pure Liquids

```
1:"A"INPUT T
3:PRINT "T=";T
5:A(32)=(T+A(2
  9))/A(27)
7:IF A(30)=0
  LET A(30)=Z
10:V=(R*A(27)/A
  (28))*A(30)^
  (1+(1-A(32))
  ^(2/7))
15:PRINT "V=";V
  ;"--CC/GMOL"
  ;"--RACKETT"
20:IF A(31)=0
  LET A(31)=(R
  *A(27)/A(28)
  )*(.292-.096
  7*W)
25:V=A(31)*(.29
  056-.08775*W
  )^((1-A(32))
  ^(2/7))
30:PRINT "V=";V
  ;"--GUNN-YAM
  ADA"
35:A(33)=1+((.7
  3098+.28908*
  W)*(1-A(32))
  )+((1.75238+
  .74293*W)*(1
  -A(32))^(1/3
  ))
40:A(34)=(Z*R*A
  (27))/A(28):
  V=A(34)/A(33
  )
45:PRINT "V=";V
  ;"--LU-ET-AL
  "
50:A(35)=.11917
  +.009153*A(3
  2)+.21091*A(
  32)^2-.06922
  *A(32)^3+.07
  48/A(32)
51:A(35)=A(35)-
  .084476*LN (
  1-A(32))
55:A(36)=.98465
  -1.60378*A(3
  2)+1.82484*A
  (32)^2-.6143
  2*A(32)^3-.3
  4546/A(32)
56:A(36)=A(36)+
  .087037*LN (
  1-A(32))
60:A(37)=-.5531
  4-.15793*A(3
  2)-1.01601*A
  (32)^2+.3409
  5*A(32)^3+.4
  6795/A(32)
61:A(37)=A(37)-
  .239938*LN (
  1-A(32))
65:A(38)=A(35)+
  A(36)*W+A(37
  )*W^2:V=A(38
  )*A(34)
70:PRINT "V=";V
  ;"--LYCKMAN-
  ET-AL"
75:END
80:"B"INPUT P
82:PRINT "P=";P
85:N=6.9547-76.
  2853*A(32)+1
  91.306*A(32)
  ^2-203.547*A
  (32)^3
86:N=N+82.7631*
  A(32)^4
90:N=(1-.89*W)*
  EXP (N)
95:A(40)=V^-1*(
  1+((9*Z*N*(P
  -A(39)))/A(2
  8))^(1/9)
100:V=1/A(40)
105:PRINT "V=";V
  ;"--CHUEH-PR
  AUSNITZ"
110:END
```

Table 2.2
Program 8
Sample Output, Input Data, Storage Information
and Program Notes

```
             "A"   T=37.78
                   V=87.06994538--C
                   C/GMOL--RACKETT
                   V=86.89834655--G
                   UNN-YAMADA
                   V=86.01721613--L
                   U-ET-AL
             "B"   V=86.80038409--L
                   YCKMAN-ET-AL
                   P=108.87
                   V=80.3007504--CH
                   UEH-PRAUSNITZ
```

For Saturated Volume "A"
    Input Data: t = 37.78°C

For Compressed Volume "B"
    Input Data: P = 108.87 atm

Storage Information

    R   = gas law constant (82.06 cc-atm/g mole-°K)

    W   = acentric factor

    Z   = Z_c, critical compressibility factor

    A(27) = T_c, °K

    A(28) = P_c, atm

    A(29) = 273.16°K (0°C)

    A(30) = Z_{RA}, scaling Z value for Rackett equation

    A(31) = V_{scr}, scaling volume for Yamada-Gunn

    A(39) = P^s, atm (required for "B" only)

Program Notes

1. The physical data for propylene were used in the example P^s = 17.2 atm and V_{scr} = 186.8. Additional data are given in the Appendix.

2. Volumes for the saturated liquids are calculated by the three given correlation without interruption.

3. The value of V_{scr} in the Gunn-Yamada correlation may be supplied by user or set equal to "0", step 20. In the latter case the value is calculated by equation 2.3. This is done in the example.

4. If Z_{RA} is unknown, Z_c is used - step 7.

5. The eitheeen constants of the correlation of Lyckman et al are given in steps 50-61 in Program A.

6. The user may enter the value of V^s to be used in the calculation of the compressed volume, otherwise the last calculated value is used.

vapor pressure, P^s. The program, defined as "B", will automatically use the V^s value predicted by the correlation of Lyckman et al., but the user may enter another value. Sample output, input data, storage information, and program notes are listed in Table 2.2.

2.2 SATURATED LIQUID MIXTURES

For liquid mixtures at the bubblepoint, the Spencer and Danner modification (1973) of the Rackett equation:

$$V_b = R[\Sigma x_i (T_{ci}/P_{ci})] Z_{RAM}^{\left[1 + (1-T_r)^{2/7}\right]} \tag{2.13}$$

where

$$Z_{RAM} = \Sigma x_i Z_{RAi} \tag{2.14}$$

was used throughout. The adjustable parameter in equation 2.13 is T_{cm}, the critical temperature of the mixture, which determines the value of the reduced temperature, T_r. In Program 9 three methods are given to predict T_{cm}. The first one was proposed by Kay (1938):

$$T_{cm} = \Sigma x_i T_{ci} \tag{2.15}$$

the second by Li (1971):

$$T_{cm} = \Sigma \phi_i T_{ci} \tag{2.16}$$

where

$$\phi_i = x_i V_{ci} / \Sigma x_i V_{ci} \tag{2.17}$$

and the third by Chueh and Prausnitz (1967):

Volumetric Relationships for Liquids

$$T_{cm} = \Sigma \Sigma \phi_i \phi_j T_{cij} \tag{2.18}$$

where

$$T_{cij} = (T_{ci} T_{cj})^{1/2} (1-k_{ij}) \tag{2.19}$$

and

$$k_{ij} = 1.0 - [(V_{ci}^{1/3} V_{cj}^{1/3})/2(V_{ci}^{1/3} + V_{cj}^{1/3})/2] \tag{2.20}$$

Table 2.3
Program 9
Determination of Volume of Saturated Liquid Mixtures

```
1:"A"INPUT X,T
3:PRINT "X1=";
  X;"T=";T
5:Y=1-X:T=T+A(
  29)
10:M=X*A(27)+Y*
  A(32):A(36)=
  T/M
15:Z=X*A(30)+Y*
  A(31):S=X*(A
  (27)/A(28))+
  Y*(A(32)/A(3
  3))
16:IF A(36)>1
  GOTO 90
17:GOSUB 100
20:PRINT "V=";V
  ;"--CC/GMOL"
  ;"--KAY-TCM"
25:A(37)=(A(34)
  *R*A(27))/A(
  28):A(38)=(A
  (35)*R*A(32)
  )/A(33)
27:C=(X*A(37))/
  (X*A(37)+Y*A
  (38))
30:D=1-C
35:M=C*A(27)+D*
  A(32):A(36)=
  T/M
40:GOSUB 100
45:PRINT "V=";V
  ;"--LI-TCM"
50:K=1-((A(37)^
  (1/3)*A(38)^
  (1/3))^.5/((
  A(37)^(1/3)+
  A(38)^(1/3))
  /2))
55:A(39)=(A(27)
  *A(32))^.5*(
  1-K)
60:M=C^2*A(27)+
  (2*C*D*A(39)
  )+D^2*A(32):
  A(36)=T/M
65:GOSUB 100
70:PRINT "V=";V
  ;"--CHEUH-PR
  AUSNITZ-TCM"
75:PRINT "ALL-M
  ETHODS-USE-M
  OD-RACKETT-E
  QN"
80:END
90:PRINT "TR>1-
  METHOD-NOT-V
  ALID"
95:GOTO 25
100:V=R*S*Z^(1+(
  1-A(36))^(2/
  7))
105:RETURN
```

34   BASIC Programs for Chemical Engineering Design

Table 2.4
Program 9
Sample Output, Input Data, Storage Information,
and Program Notes

```
                                    X1=0.7T=37.73
                                    TR>1-METHOD-NOT-
         X1=0.3069T=37.73           VALID
         V=89.04758421--C           V=84.59851109--L
         C/GMOL--KAY-TCM            I-TCM
         V=88.31396319--L           V=93.59236491--C
         I-TCM                      HEUH-PRAUSNITZ-T
         V=88.62031485--C           CM
         HEUH-PRAUSNITZ-T           ALL-METHODS-USE-
         CM                         MOD-RACKETT-EQN
         ALL-METHODS-USE-
         MOD-RACKETT-EQN


              "A"                         "B"
```

Input Data: "A" x_i = .3069 mole fraction propane (with benzene)

              "B" x_i = .7 mole fraction methane (with n-pentane)

              $t = 37.73°C$

Storage Data:

  R   = gas law constant (82.06 cc-atm/g mole-$°K$)

  A(27) = T_{c1}, $°K$

  A(28) = P_{c1}, atm

  A(29) = 273.16$°K$ (0$°C$)

  A(30) = Z_{RA1}, scaling Z value for Rackett equation (comp 1)

  A(31) = Z_{RA2}, scaling Z value for Rackett equation (comp 2)

  A(32) = T_{c2}, $°K$

  A(33) = P_{c2}, atm

  A(34) = Z_{c1}, critical compressibility factor (comp 1)

  A(35) = Z_{c2}, critical compressibility factor (comp 2)

Program Notes:

  1. The method Kay predicts the lowest T_{cm} value, hence most frequently cannot be used. See output "B".

  2. Critical volumes are calculated by $V_c = Z_c RT_c/P_c$ (step 25). A(37) = V_{c1} and A(38) = V_{c2}.

Volumetric Relationships for Liquids 35

of the three methods, the value of T_{cm} predicted by the method of Li is the highest and seems to most nearly approach the actual critical temperature of the mixture. The method of Kay predicts the lowest value of T_{cm}, hence may be considerably less than the real value of T_c for mixture. This means that when Kay's method is used that the calculated T_r used in equation 2.13 may be greater than 1.0 which gives erroneous answers. Allowance is made for this in the program - step 16.

A test is made of the value of T_r and if pertinent the message "$T_r > 1$ - method not valid" is printed out, but calculations continue. Program 9 is given in Table 2.3 and sample output, input data, storage information, and program notes in Table 2.4.

3
Vapor Pressure — Temperature Correlations

The vapor pressure-temperature relationship of fluids is perhaps the most frequently investigated physical phenomenon. Many empirical expressions of this relationship have been developed. In this section programs are given for four commonly used expressions - Lee and Kesler (1975), Antoine (1888), Riedel-Plank-Miller (1977) and Frost-Kalkwarf-Thodos (1953). With the exception of the Antoine equation, which is usually employed at pressures up to two atmospheres, the expressions are valid to the critical point. In all cases programs are given for the determination of the vapor pressure given the temperature and for the temperature given the pressure.

3.1 LEE-KESLER CORRELATION

The Lee-Kesler relationship is:

$$\ln P_r^s = f^{(0)}(T_r) + \omega f^{(1)}(T_r) \qquad (3.1)$$

where

$$f^{(0)} = 5.92714 - 6.09648/T_r - 1.28862 \ln T_r + 0.16934 \, T_r^6 \quad (3.2)$$

$$f^{(1)} = 15.2518 - 15.6875/T_r - 13.4721 \ln T_r + 0.43577 \, T_r^6 \quad (3.3)$$

and the acentric factor, ω, is calculated by

$$\omega = (-\ln P_c - (f_b^{(0)}))/f_b^{(1)} \quad (3.4)$$

where $f_b^{(0)} = f^{(0)}$ in equation 3.2 and $f_b^{(1)} = f^{(1)}$ in equation 3.3 with $T_r = T_{br}$ and $P = 1$ atm in equation 3.1. The program defined as "A", Program 10 - Table 3.1, is for the solution for pressure where $t(^\circ C)$ is given. The pressure can be calculated directly and its value is printed. Sample output and other information are given in Table 3.2.

The program defined as "B", Table 3.1, is for the solution of the temperature, given the pressure. This solution is more complex than the previous one and the Newton technique is used to converge on the correct temperature. The normal boiling conditions (t and P) are used as the initial estimate. In step 60 the new estimate of the temperature is determined by:

$$T1 = T + (P-P_{cal})/(P_c(dP_r/dT_r)) \quad (3.5)$$

(dP_r/dT_r) is determined from equation 2.1 in steps 50 and 55. The derivative is labelled as D. When the difference between the calculated pressure and the given pressure is less than some arbitrarily defined quantity--.005P in this case--the solution is considered satisfactory and the temperature value and the calculated pressure are printed. Sample output is given in Table 3.2 with input data, storage information and program notes.

Vapor Pressure-Temperature Correlations

Table 3.1
Program 10
Lee-Kesler and Antoine Relationships

```
1:"A"INPUT T
2:PRINT "T=";T
3:X=0
5:IF W=0GOTO 1
  00
10:A(30)=(T+A(2
  9))/A(27)
15:GOSUB 150
20:P=F+(W*G)
25:P=EXP (P)*A(
  28)
30:PRINT "P=";P
  ;"--ATM"
31:PRINT "LEE-K
  ESLER"
32:END
35:"B"INPUT P
36:PRINT "P=";P
37:X=1
38:A(36)=P/A(28
  )
40:T=A(31):A(32
  )=1:A(30)=T/
  A(27)
45:IF W=0GOTO 1
  00
50:D=6.09648/A(
  30)^2-1.2886
  2/A(30)+(6*.
  169347*A(30)
  ^5)
55:D=(D+W*(15.6
  875/A(30)^2-
  13.4721/A(30
  )+(6*.43577*
  A(30)^5))*A
  (36)
60:A(33)=A(30)+
  ((P-A(32))/(
  A(28)*D))
65:IF ABS (P-A(
  32))<.005*P
  GOTO 80
66:A(34)=A(33)*
  A(27)
67:PAUSE USING
  ;A(34)
70:A(30)=A(33)
72:GOSUB 150
75:A(35)=F+(W*G
  ):A(32)=EXP
  (A(35))*A(28
  )
77:GOTO 50
80:T=(A(30)*A(2
  7))-A(29)
82:PRINT "T=";T
  ;"---PCALD="
  ;A(32)
83:PRINT "LEE-K
  ESLER"
85:END
100:A(30)=A(31)/
  A(27)
105:GOSUB 150
110:W=(-(LN (A(2
  8))+F))/G
112:IF X=0GOTO 1
  0
113:IF X=1GOTO 5
  0
150:F=5.92714-(6
  .09648/A(30)
  )-(1.28862*
  LN (A(30)))+
  (.16934*A(30
  )^6)
155:G=15.2518-(1
  5.6875/A(30)
  )-(13.4721*
  LN (A(30)))+
  (.43577*A(30
  )^6)
160:RETURN
170:"C"INPUT T
172:PRINT "T=";T
175:P=A-(B/(T+C)
  ):P=EXP (2.3
  0258*P)
177:PRINT "P=";P
  ;"--MM-HG"
178:PRINT "ANTOI
  NE"
179:END
180:"D"INPUT P
182:PRINT "P=";P
185:T=(B/(A-LOG
  (P)))-C
186:PRINT "T=";T
  ;"--CENTI"
187:PRINT "ANTOI
  NE"
188:END
```

Table 3.2
Program 10
Sample Output, Input Data, Storage Information
and Program Notes

"A" and "B"
```
T=15.56
P=7.362668364--A
TM
LEE-KESLER
P=7.321
T=15.34668658---
PCALD=7.32005367
4
LEE-KESLER
```

"C" and "D"
```
T=-42.07
P=759.9918728--M
M-HG
ANTOINE
P=760.
T=-42.07009011--
CENTI
ANTOINE
```

Input Data: "A" $t = 15.56\,°C$

"B" $P = 7.321$ atm

"C" $t = -42.07\,°C$ (normal boiling temperature)

"D" $P = 760$ mmHg

Storage Information:

  A

  B   Antoine Constants

  C

  w   = acentric factor

  $A(27) = T_c$, $°K$

  $A(28) = P_c$, atm

  $A(29) = 273.16\,°K$ $(0\,°C)$

  $A(31) = T_b$, $°K$

Program Notes:

1. Physical data for propane were used in the example. A = 6.82973, B = 813.2 and C = 248. Additional data are included in the Appendix.

2. The value of the acentric factor may be entered by the user, if not let w = 0 and the acentric factor is calculated by equation 3.4. Subroutine 100 in the program. In the example, the literature value of the acentric factor was used.

3. In "A" the value of x is set at "0" or "1" (steps 3 and 37) to give the proper direction to the calculations (see steps 112 and 113).

Vapor Pressure-Temperature Correlations 41

3.2 ANTOINE EQUATION

The Antoine equation is commonly used to fit low pressure-temperature data. The equation is:

$$\log P^S = A - B/(t + C) \qquad (3.6)$$

and is explicit in both pressure and temperature, so direct solutions are possible. These are given in Program 10 (Table 3.1) and sample output, storage information, and program notes are given in Table 3.2. The program for the solution of pressure given temperature is defined "C" and the other option defined "D".

3.3 RIEDEL, PLANK AND MILLER CORRELATION

The Riedel, Plank and Miller correlation is another expression for the reduced vapor pressure and is explicit in P_r, implicit in T_r and is:

$$\ln P_r^S = -G/T_r [1 - T_r^2 + k(3 + T_r)(1 - T_r)^3] \qquad (3.7)$$

where

$$G = 0.4835 + 0.4605h \qquad (3.8)$$

$$k = [h/G - (1 + T_{br}) / (3 + T_{br})(1 - T_{br})^2] \qquad (3.9)$$

$$h = T_{br}(\ln P_c)/(1 - T_{br}) \qquad (3.10)$$

Program 11 given in Table 3.3 presents the two solutions for this relationship. The pressure, given temperature, can be determined directly. The program is defined as "A". The program defined as "B" is the solution of the temperature given the vapor pressure. A trial and error solution using the Newton technique is used - steps 50-65. The calculated reduced pressure is compared to an assumed

Table 3.3
Program 11
Riedel-Planck-Miller and
Frost-Kalkwarf-Thodos Correlations

```
1:"A"INPUT T
2:PRINT "T=";T
5:A(30)=(T+A(2
   9))/A(27)
7:A(37)=A(31)/
   A(27)
10:H=(A(37)*LN
   (A(28)))/(1-
   A(37))
12:G=.4835+.460
   6*H
15:K=((H/G)-(1+
   A(37)))/((3+
   A(37))*(1-A(
   37))^2)
20:P=(-G/A(30))
   *(1-A(30)^2+
   (K*(3+A(30))
   *(1-A(30))^3
   )):P=EXP (P)
   *A(28)
25:PRINT "P=";P
   ;"--ATM"
26:PRINT "RIEDE
   L-PLANK-MILL
   ER"
27:END
30:"B"INPUT P
32:PRINT "P=";P
35:A(37)=A(31)/
   A(27)
37:H=(A(37)*LN
   (A(28)))/(1-
   A(37))
40:G=.4835+.460
   6*H
42:K=((H/G)-(1+
   A(37)))/((3+
   A(37))*(1-A(
   37))^2)
45:A(30)=A(37):
   A(36)=P/A(28
   )
50:A(35)=(-G/A(
   30))*(1-A(30
   )^2+(K*(3+A(
   30))*(1-A(30
   ))^3)):A(35)
   =EXP (A(35))
55:D=(-G*A(36))
   *(-A(30)^-2-
   1+(K*(-3*A(3
   0)^-3+6-3*A(
   30)^2)))
60:A(33)=A(30)-
   ((A(35)-A(36
   ))/D)
65:IF ABS (A(35
   )-A(36))<.00
   05*A(36)GOTO
   100
67:T=A(33)*A(27
   )
70:PAUSE USING
   ;T
72:A(30)=A(33)
73:PAUSE USING
   ;A(30)
75:GOTO 50
100:P=A(35)*A(28
   ):T=(A(30)*A
   (27))-A(29)
105:PRINT "PCALD
   =";P;"T=";T
108:PRINT "RIEDE
   L-PLANK-MILL
   ER"
110:END
120:"C"INPUT T
122:PRINT "T=";T
125:T=T+A(29)
130:A(35)=A+B/T+
   C*LN (T):A(3
   5)=EXP (A(35
   ))
135:P=A+B/T+C*LN
   (T)+((D*A(35
   ))/T^2):P=
   EXP (P)
140:IF ABS (P-A(
   35))<.0005*P
   GOTO 160
142:PAUSE USING
   ;P
145:A(35)=P
150:GOTO 135
160:P=P/760
165:PRINT "P=";P
   ;"--ATM"
168:PRINT "FROST
   -KALKWARF"
170:END
180:"D"INPUT P
182:PRINT "P=";P
185:P=P*760:A(34
   )=A(29)
190:T=(-B+(ABS (
   B)^2-(4*(A+C
   *LN (A(34))-
   LN (P))*D*P)
   )^.5)/(2*(A+
   C*LN (A(34))
   -LN (P)))
193:IF ABS (A(34
   )-T)<.05GOTO
   210
195:PAUSE USING
   ;T
198:A(34)=T
200:GOTO 190
210:P=P/760:T=T-
   A(29)
213:PRINT "T=";T
   ;"--CENTI"
214:PRINT "FROST
   -KALKWARF"
215:END
```

Vapor Pressure-Temperature Correlations 43

Table 3.4
Program 11
Sample Output, Input Data,
Storage Information and Program Notes

```
T=15.56
P=7.256453972--A
TM
RIEDEL-PLANK-MIL
LER
P=7.321
PCALD=7.31855268
4T=15.87567434
RIEDEL-PLANK-MIL
LER


T=15.56
P=7.301003407--A
TM
FROST-KALKWARF
P=7.321
T=15.6425814--CE
NTI
FROST-KALKWARF
```

Input Data: "A" $t = 15.56°C$

"B" $p = 7.321$ atm

"C" $t = 15.56°C$

"D" $p = 7.321$ atm

Storage Information:

A
B   Constants for Frost-Kalkwarf-Thodos eqn.
C
D

$A(27) = T_c,\ °K$

$A(28) = P_c,\ atm$

$A(29) = 273.16°K\ (0°C)$

$A(31) = T_b,\ °K$

Programs:

1. Propane data were used in the analysis A = 43.492, B = -326.92, C = -4.179 and D = 1.81 (Harlacher and Braun (1870)).

value within an arbitrarily selected limit - .0005 P_r^s. When this condition is satisfied, the calculated pressure and temperature are printed. Output from these two programs, input data, storage information, and program notes are given in Table 3.4.

3.4 FROST-KALKWARF-THODOS CORRELATION

Also, in Program 11 are the solutions for the Frost-Kalkwarf-Thodos equation:

$$\ln P^s = A + B/T + C \ln T + D P^s/T^2 \tag{3.11}$$

This equation involves a trial and error solution for either vapor pressure or temperature. If the temperature is known, an estimated value of pressure may be determined by neglecting the last term in equation 3.11 - step 130. The estimated value is then used to determine P^s by using all terms in the equation - step 135. The value of P^s then becomes the estimated value and the calculation repeated until the two pressure values check within $\pm$ .0005 P - or any limit selected by the user. This program is defined as "C".

The program defined as "D" is for the solution of the temperature. When the vapor pressure is given, equation 11 reduces to a quadratic with the exception of the ln T term which is relatively minor. Hence, as a first estimated T is set at 273.16°K on the RHS of the relationship:

$$T = \frac{(-B + (B^2 - (4(A + \ln C\, T_{est} - \ln P)DP))^{1/2})}{2(A + C \ln T_{est} - \ln P)} \tag{3.12}$$

The temperature value determined by this equation, step 190, is then used as the T_{est} and calculations proceed until the absolute difference between the temperature values is less than .05 - step 193. Sample output, input data, storage information, and program notes for both programs are given in Table 3.4.

4
Latent Heat of Vaporization

The latent heat of vaporization of a pure substance may be calculated from empirical correlations, or provided suffice data are available, by thermodynamic relationships. In the former category are the Riedel expression (1954) for the latent heat of vaporization at the normal boiling point, a simple relationship proposed by Watson (1943) to evaluate the effect at temperature on the latent heat, and the generalized relationship proposed by Pitzer et al. (1955). Alternately the latent heat of vaporization may be calculated by the Clapeyron equation:

$$dP/dT = \Delta H_v / T \Delta V_v \qquad (4.1)$$

which is thermodynamically rigorous. To use equation 1 requires knowledge of the vapor pressure-temperature relationship and volumetric data for the saturated liquid and vapor-phases. Both types of solutions are included in Program 12.

4.1 EMPIRICAL CORRELATIONS

The Riedel expression:

$$\Delta H_{vb} = 1.093 \, RT_c (T_{br}((\ln P_c - 1)/(.930 - T_{br}))) \qquad (4.2)$$

Table 4.1
Program 12
Calculation of Latent Heat of Vaporization

```
1:"A"INPUT T
3:PRINT "TB=";
  T
5:A(31)=T+A(29
  )
8:A(32)=A(31)/
  A(27)
10:H=1.093*A(27
   )*R*((A(32)*
   (LN (A(28))-
   1))/(.930-A(
   32))):H=H*24
   .218
11:A(30)=H
12:PRINT "HVB="
   ;H;"--CAL/GM
   OL";"--RIEDE
   L"
15:END
20:"B"INPUT T
22:PRINT "T=";T
25:A(33)=(T+A(2
   9))/A(27)
30:H=A(30)*((1-
   A(33))/(1-A(
   32)))^.378
35:PRINT "H=";H
   ;"--WATSON"
37:END
40:"C"INPUT T
42:PRINT "T=";T
45:T=(T+A(29))/
   A(27)
50:H=R*A(27)*((
   7.08*(1-T)^.
   354)+(10.95*
   W*(1-T)^.456
   ))*24.218
55:PRINT "H=";H
   ;"--PITZER-E
   T-AL"
60:END
65:"D"INPUT T
66:PRINT "T=";T
67:T=(T+A(29))/
   A(27)
70:A(34)=6.0964
   8-1.28862*T+
   1.016*T^7+(W
   *(15.6875-13
   .4721*T+2.61
   5*T^7))
75:F=5.92714-6.
   09648/T-1.28
   862*LN (T)+.
   169347*T^6
80:G=15.2518-15
   .6875/T-13.4
   721*LN (T)+.
   43577*T^6
85:P=F+W*G:P=
   EXP (P)*A(28
   )
87:PRINT "P=";P
90:A(36)=(R*A(2
   7)/A(28))*A(
   35)^(1+(1-T)
   ^(2/7))
92:PRINT "VL=";
   A(36)
95:K=.37464+1.5
   4226*W-.2699
   2*W^2
100:Q=1+(K*(1-T^
    .5)):Q=Q^2
110:A=(.45724*Q*
    P)/(A(28)*T^
    2):B=(.0778*
    P)/(A(28)*T)
120:X=(A*B-B^2-B
    ^3):Y=(A-3*B
    ^2-2*B):Z=1
125:G=Z^3-((1-B)
    *Z^2)+Y*Z-X
130:I=3*Z^2-(2*(
    1-B)*Z+Y
135:A(37)=Z-(G/I
    )
140:J=A(37)^3-((
    1-B)*A(37)^2
    )+Y*A(37)-X
145:IF ABS (J)<.
    00001GOTO 18
    0
150:Z=A(37)
151:PAUSE USING
    ;Z
155:GOTO 125
180:Z=A(37):V=(Z
    *R*(T*A(27))
    /P
185:PRINT "Z=";Z
    ;"V=";V
190:H=(A(34)*(V-
    A(36))*P*24.
    219)/T
195:PRINT "H=";H
    ;"--CAL/GMOL
    "
```

Latent Heat of Vaporization

requires only the knowledge of the normal boiling point and the critical temperature and pressure. The latent heat of vaporization at the normal boiling point can then be determined. The solution to equation 4.2 is defined as "A" in the Program 12, Table 4.1.

Once a value of the latent heat at a given temperature is known, the Watson correlation:

$$\Delta H_{v(t2)} = \Delta H_{v(t1)} ((1 - T_{r2})/(1 - T_{r1}))^{0.378} \qquad (4.3)$$

can be used to evaluate this heat effect at a second temperature. No additional data are required. The program is set up so that the value obtained from the Riedel correlation is used. However, the user has the option to substitute another value. The user may wish to use a latent heat value other than at the normal boiling. This can be done, but both the latent heat and reduced temperature values must be put in the proper storage location--A(30) and A(32) respectively. This program is defined as "B".

The last empirical relationship programmed is that of Pitzer et al. which was extended by Carruth and Kobayashi (1972) and put in following mathematical form by Reid, Prausnitz, and Sherwood (1977):

$$\Delta H_v / RT_c = 7.08(1 - T_r)^{0.354} + 10.95\omega(1 - T_r)^{0.456} \qquad (4.4)$$

This relationship is reasonably accurate ($\pm$ 2%) and requires a minimum of data--the critical temperature and the acentric factor. The program is defined as "C".

4.2 CLAPEYRON EQUATION

Equation 4.1 may be written in reduced form:

$$-d \ln P_r^s / d(1/T_r) = \Delta H_v T_r / \Delta V_v P_v^s \qquad (4.5)$$

Table 4.2
Program 12
Sample Output, Input Data,
Storage Information, and Program Notes

```
TB=-161.45
HVB=1986.565329-
-CAL/GMOL--RIEDE
L
T=-156.71
H=1940.575611--W
ATSON
T=-156.71
H=1941.488081--P
ITZER-ET-AL
T=-156.71
P=1.459116313
VL=3.868435046E-
02
Z=9.562934256E-0
1V=6.262853641
H=1944.678631--C
AL/GMOL
```

Input Data: "A" t = -161.45°C normal boiling temperature of methane

"B" t = -156.71°C

"C" t = -156.71°C

"D" t = -156.71°C

Storage Information:

R = gas law constant (0.08206 liter-atm/g mole °K)

w = acentric factor

$A(27) = T_c$, °K

$A(28) = P_c$, atm

$A(29) = 273.16$°K (0°C)

$A(35) = Z_{RA}$, modified Z_c for Rackett correlation

Program Notes:

1. Methane data were used in the example.

Latent Heat of Vaporization

In Program 12, the necessary vapor pressure-temperature data were evaluated by the Lee-Kesler correlation-equation 3.1. The slope term from the Lee-Kesler relationship is:

$$-d \ln P_r^s/d(1/T_r) = 6.0948 - 1.28862T_r + 1.016T_r^7$$
$$+ \omega(15.6875 - 13.4721T_r + 2.615T_r^7) \tag{4.6}$$

Other relationships given in Chapter 3 may be used in place of that of Lee-Kesler. For ΔV_v term, the Peng Robinson EOS--equation 1.27-- was used to evaluate the saturated vapor volume and the modified Rackett relationship--equation 2.1--the saturated liquid volume. This program is defined as "D".

Program 12 is given in Table 4.1. Sample output, input data storage information, and program notes are given in Table 4.2.

5
Heat Capacity Expressions for Gases

The heat capacity of gases at isobaric conditions is usually expressed as a function of temperature in the forms:

$$C_p = a + bT + cT^2 + dT^3 \qquad (5.1)$$

or

$$C_p = a + bT + c/T^2 \qquad (5.2)$$

The available data are largely for gases in the ideal gas state. From these equations ΔH (or Q_p) and ΔS can be calculated provided the initial and final temperatures are given. This is a straight forward solution. Or, given an initial temperature, and the value of ΔH, or ΔS, the final temperature can be calculated by a trial and error solution. In this section, programs to treat both types of problems for the two thermodynamic functions are offered.

5.1 $C_p = a + bT + cT^2 + dT^3$

In program 13 the four solutions using equation 1 are given. The first program, defined as "A", solves for ΔH given the initial and final temperature. The basic relationship used is:

$$\Delta H = Q_p = \int_{T_1}^{T_2} C_p \, dT \qquad (5.3)$$

And, for equation 1,

$$\Delta H = Q_p = a(T_2 - T_1) + b/2(T_2^2 - T_1^2) + c/3(T_2^3 - T_1^3) + d/4(T_2^4 - T_1^4) \qquad (5.4)$$

The answer in cal/g mol is printed.

The next program, defined as "B", solves for the final temperature given the initial temperature and the value of ΔH. The given values and an estimate of the final temperature are entered. The Newton technique is used to converge on the correct final temperature. The equation used is:

$$T_3 = T_2 - (Q2 - Q)/f'(x) \qquad (5.5)$$

Q is the given ΔH, and Q2 is determined using the estimated temperature, T_2, and $f'(x)$ is the derivative, $d(Q2 - Q)/dT$. T_3 is the new estimated temperature value. In the program, this is step 47 and the derivative is calculated in step 35. When the absolute difference between (Q2 - Q) is less than 0.0005Q, an arbitrarily selected quantity, the estimated temperature value is satisfactory and its value is printed.

Next part of Program 13, defined as "C", is for the determination of ΔS given the initial and final temperature by:

$$\Delta S = a \ln(T_2/T_1) + b(T_2 - T_1) + c/2(T_2^2 - T_1^2) + d/3(T_2^2 - T_1^3) \qquad (5.6)$$

This is a straight forward solution and the ΔS value in cal/g mol - °K is printed.

Heat Capacity Expressions for Gases   53

Table 5.1
Program 13
Calculation of Thermodynamic Properties Using
Heat Capacity Expressions: $a + bT + cT^2 + dT^3$

```
 1:"A"INPUT A(5
   ),T
 3:PRINT "T1=";
   A(5);"T2=";T
 5:A(5)=A(5)+A(
   29):T=T+A(29
   )
10:H=A*(T-A(5))
   +(B*(T^2-A(5
   )^2))/2+(C*(
   T^3-A(5)^3))
   /3+(D*(T^4-A
   (5)^4))/4
15:PRINT "H=";H
   ;"--CAL/GMOL
   "
17:END
20:"B"INPUT A(5
   ),Q,T
22:PRINT "T1=";
   A(5);"H=";Q
25:A(5)=A(5)+A(
   29):T=T+A(29
   )
30:A(6)=A*(T-A(
   5))+(B*(T^2-
   A(5)^2))/2+(
   C*(T^3-A(5)^
   3))/3+(D*(T^
   4-A(5)^4))/4
35:A(7)=A+(B*T)
   /2+(C*T^2)/3
   +(D*T^3)/4
40:A(8)=T-((A(6
   )-Q)/A(7))
42:PAUSE USING
   ;A(8)
45:IF ABS (A(6)
   -Q)<.0005*Q
   GOTO 65
47:T=A(8)
50:GOTO 30
65:T=A(8)-A(29)
70:PRINT "T2=";
   T
75:END
80:"C"INPUT A(5
   ),T
82:PRINT "T1=";
   A(5);"T2=";T
85:A(5)=A(5)+A(
   29):T=T+A(29
   )
90:S=A*LN (T/A(
   5))+B*(T-A(5
   ))+(C*(T^2-A
   (5)^2)/2+(D
   *(T^3-A(5)^3
   ))/3
95:PRINT "S=";S
   ;"--CAL/GMOL
   -K"
97:END
100:"D"INPUT A(5
   ),S,T
102:PRINT "T1=";
   A(5);"S=";S
105:A(5)=A(5)+A(
   29):T=T+A(29
   )
110:A(6)=A*LN (T
   /A(5))+B*(T-
   A(5))+(C*(T^
   2-A(5)^2))/2
   +(D*(T^3-A(5
   )^3))/3
115:A(7)=A/T+B+C
   *T+D*T^2
120:A(8)=T-((A(6
   )-S)/A(7))
122:PAUSE USING
   ;A(8)
125:IF ABS (A(6)
   -S)<.0005*S
   GOTO 140
130:T=A(8)
135:GOTO 110
140:T=A(8)-A(29)
145:T=A(8)-A(29)
150:PRINT "T2=";
   T
152:END
```

Table 5.2
Program 3
Sample Output, Input Data,
Storage Information and Program Notes

```
T1=50.T2=100.
H=348.5360872--C
AL/GMOL
T1=50.H=348.54
T2=100.0003386
T1=50.T2=100.
S=1.002788949--C
AL/GMOL-K
T1=50.S=1.0028
T2=100.000591
```

Input Data: "A" $t_1 = 50°C$ and $t_2 = 100°C$

"B" $t_1 = 50°C$, $\Delta H = 348.54$ cal/g mol, and t_2 est($°C$), t_2 est($°C$)

"C" $t_1 = 50°C$ and $t_2 = 100°C$

"D" $t_1 = 50°C$ $\Delta S = 1.0028$ cal/g mole $- °K$, and t_2 est($°C$)

Storage Data:

   A

   B

   C   Constants of Equation 5.1

   D

   $A(29) = 273.16°K$ $(0°C)$

Programs Notes:

1. Nitrogen data were used in the examples. $A = 7.44$; $B = -3.24 \times 10^{-3}$; $C = 6.4 \times 10^{-6}$; $D = 2.76 \times 10^{-9}$ (Prausnitz et al. (1977)).

The final program, defined as "D", solves for the final temperatures given the initial temperature, ΔS, and are estimated final temperature. The technique is the same as used in "B" and involves steps 110 to 120 in the program.

Similar to the previous calculation, when the absolute difference between ($S_{calc} - S$) is less than $0.0005S$, the estimated final temperature value is considered satisfactory and is printed.

Program 13 is given in Table 5.1 and sample output, input data, storage information are given in Table 5.2.

Table 5.3
Program 14
Calculation of Thermodynamic Properties Using
Heat Capacity Expression: $C_p = a + bT + c/T^2$

```
 1:"A"INPUT A(5
   ),T
 2:PRINT "T1=";
   A(5);"T2=";T
 5:A(5)=A(5)+A(
   29)
 6:T=T+A(29)
 7:Q=A*(T-A(5))
   +(B*(T^2-A(5
   )^2))/2-C*(1
   /T-1/A(5))
 8:PRINT "Q=";Q
   ;"--CAL/GMOL
   "
 9:END
10:"B"INPUT A(5
   ),Q,T
11:PRINT "T1=";
   A(5);"Q=";Q
12:A(5)=A(5)+A(
   29)
13:T=T+A(29)
14:A(6)=A*(T-A(
   5))+(B*(T^2-
   A(5)^2))/2-C
   *(1/T-1/A(5)
   )
15:A(7)=A+(B*T)
   /2+C/T^2
16:A(8)=T-((A(6
   )-Q)/A(7))
17:PAUSE USING
   ;A(8)
19:IF ABS (A(6)
   -Q)<.0005*Q
   GOTO 25
20:T=A(8)
21:GOTO 14
25:T=T-A(29)
26:PRINT "T2=";
   T
29:END
30:"C"INPUT A(5
   ),T
31:PRINT "T1=";
   A(5);"T2=";T
32:A(5)=A(5)+A(
   29)
33:T=T+A(29)
34:S=A*LN (T/A(
   5))+B*(T-A(5
   ))-(C*(1/T^2
   -1/A(5)^2))/
   2
35:PRINT "S=";S
   ;"--CAL/GMOL
   -K"
36:END
40:"D"INPUT A(5
   ),S,T
41:PRINT "T1=";
   A(5);"S=";S
42:A(5)=A(5)+A(
   29)
43:T=T+A(29)
44:A(6)=A*LN (T
   /A(5))+B*(T-
   A(5))-(C*(1/
   T^2-1/A(5)^2
   ))/2
45:A(7)=A/T+B+C
   /T^3
46:A(8)=T-(A(6)
   -S)/A(7)
47:PAUSE USING
   ;A(8)
49:IF ABS (A(6)
   -S)<.0005*S
   GOTO 58
51:T=A(8)
53:GOTO 44
58:T=A(8)-A(29)
60:PRINT "T2=";
   T
61:END
```

5.2 $C_p = a + bT + c/T^2$

Program 14 is similar to the previous program with the difference that equation 2 is used throughout. Thus the calculation of the enthalpy change becomes:

$$\Delta H = Q_p = a(T_2 - T_1) + b/2(T_2^2 - T_1^2) - c(1/T_2 - 1/T_1) \qquad (5.7)$$

The program, defined as "A", solves for the enthalpy change (Q_p) given the initial and final temperatures. The Q value is printed.

Table 5.4
Program 14
Sample Output, Input Data,
Storage Information, and Program Notes

```
         "A"    T1=50.T2=100.
                Q=352.191679--CA
                L/GMOL
         "B"    T1=50.Q=352.19
                T2=100.0003787
         "C"    T1=50.T2=100.
                S=1.013196954--C
                AL/GMOL-K
         "D"    T1=50.S=1.0132
                T2=100.0001605
```

Input Data: "A" $t_1 = 50°C$ and $t_2 = 100°C$
             "B" $t_1 = 50°C$, $\Delta H = 352.19$ cal/g mole, and $t_2$est($°C$)
             "C" $t_1 = 50°C$ and $t_2 = 100°C$
             "D" $t_1 = 50°C$, $\Delta S = 1.0132$ cal/g mol $°K$, and
                  t_2 est($°C$)

Storage Information:
    A
    B   Constants of Equation 5.2
    C
    A(29) = 273.16°K (0°C)

Program Notes:
1. Nitrogen data were used in the example. A = 6.83; B = .009;
    C = -12000 (Kelly (1960)).

Heat Capacity Expressions for Gases   57

In the program, defined as "B", the final temperature is calculated, given the initial temperature and the enthalpy change. An estimated value of the final temperature is also entered. This requires a trial and error solution and, as previously, the Newton technique is used--equation 5.5--to obtain the correct final temperature. These are steps 14-16. When the calculated enthalpy and the given value differ by less than 0.0005Q, the final temperature is assumed to be correct.

The next program, defined as "C", solves for the change in entropy, ΔS, given the initial and final temperatures. The expression used is

$$\Delta S = a \ln(T_2/T_1) + b(T_2 - T_1) - c/2(1/T_2^2 - 1/T_1^2) \qquad (5.8)$$

The value of the entropy change is printed.

The final program, defined as "D", solves for the final temperature, when the initial temperature, ΔS, and an estimated value of the final temperature are entered. This is a trial and error solution and the Newton technique is used--steps 44-47. Where a agreement between the calculated values of the entropy change and the given is satisfactory--step 49--the final temperature is printed.

Program 14 is given in Table 5.3 and sample output, input data, storage information, and program notes are given in Table 5.4.

6
Calculation of Total Changes in Enthalpy and Entropy

Programs 1-6 permit the calculation of changes in enthalpy and entropy under isothermal conditions and programs 13 and 14 provide means to evaluate changes in these thermodynamic functions under isobaric conditions. In this chapter, programs are given for the determination of total changes in enthalpy and entropy and the special case of the isentropic process, i.e., $\Delta S = 0$. The basic relationships used are:

$$dH = (\partial H/\partial T)_p dT + (\partial H/\partial P)_T dP \qquad (6.1)$$

and

$$dS = (\partial S/\partial T)_p dT + (\partial S/\partial P)_T dP \qquad (6.2)$$

In both instances the evaluation of the first partial change requires heat capacity information and the second partial change, P-V-T information; e.g. an equation of state.

6.1 CALCULATION OF ΔH AND ΔS

In Program 15, elements of Programs 1 and 13 are combined to permit the calculation of ΔH and ΔS given the initial and final temperatures and pressures. The calculation follow a three step procedure:

1. Calculation of the ΔH1 (ΔS1) from the initial conditions of temperature (t_1 and pressure P_1) to the ideal gas state at the t_1 and a reference pressure--assumed to be one atmosphere. The assumed value of P_{ref} is unimportant since it cancells out.
2. Calculation of ΔH_2 (or ΔS_2) from the initial temperature to the final temperature (t_2) in the ideal gas state.
3. Calculation of ΔH_3 (or ΔS_3) from the final temperature and the ideal gas state to the real gas state at the final pressure (P_2) and t_2. The total change in enthalpy (ΔH) is:

$$\Delta H = \Delta H_3 + \Delta H_2 - \Delta H_1 \tag{6.3}$$

similarly:

$$\Delta S = \Delta S_3 + \Delta S_2 - \Delta S_1 \tag{6.4}$$

The Redlich-Kwong EOS was used to calculate the initial and final volumes (V_1 and V_2) and, in turn, ΔH_1 and ΔH_3. The procedure is the same as used in Program 1. ΔH_2 was calculated using the heat capacity expression equation 5.1. This step utilizes portions of Program 13. Program 15 is given in Table 6.1 and sample output, input data, storage information and program notes are given in Table 6.2.

Table 6.1
Program 15
Calculation of Total Changes in Enthalpy and Entropy

```
1:"A"INPUT A(5
 ),T,A(6),P
3:PRINT "T1=";
 A(5);"T2=";T
5:PRINT "P1=";
 A(6);"P2=";P
7:A(5)=A(5)+A(
 29):T=T+A(29
 )
8:A(33)=P:A(34
 )=T
10:A(31)=A*(T-A
 (5))+(B*(T^2
 -A(5)^2))/2+
 (C*(T^3-A(5)
 ^3))/3+(D*(T
 ^4-A(5)^4))/
 4
11:A(41)=A*LN (
 T/A(5))+B*(T
 -A(5))+(C*(T
 ^2-A(5)^2))/
 2+(D*(T^3-A(
 5)^3))/3
12:PRINT "H2=";
 A(31);"--CAL
 /GMOL"
13:PRINT "S2=";
 A(41);"--CAL
 /GMOL-K"
15:T=A(5):P=A(6
 )
18:GOSUB 200
20:A(30)=((((3*
 A(35))/(2*A(
 36)*T^.5))*
 LN (M))-(R*T
 )+(P*V))*24.
 2179
21:GOSUB 300
22:PRINT "S1=";
 S
23:A(40)=S
25:PRINT "Z1=";
 A(39);"V1=";
 V;"H1=";A(30
 )
30:T=A(34):P=A(
 33)
35:GOSUB 200
40:A(32)=((((3*
 A(35))/(2*A(
 36)*T^.5))*
 LN (M))-(R*T
 )+(P*V))*24.
 2179
41:GOSUB 300
42:A(42)=S
45:PRINT "Z2=";
 A(39);"V2=";
 V;"H3=";A(32
 )
46:PRINT "S3=";
 S
50:H=A(32)+A(31
 )-A(30)
51:S=A(42)+A(41
 )-A(40)
55:PRINT "H=";H
 ;"--CAL/GMOL
 "
56:PRINT "S=";S
 ;"--CAL/GMOL
 -K"
60:END
200:A(35)=(.4278
 *R^2*A(27)^2
 .5)/A(28):A(
 36)=(.0867*R
 *A(27))/A(28
 )
205:A(37)=A(35)/
 (R^2*T^2.5):
 A(38)=A(36)/
 (R*T)
210:X=(A(38)*P+(
 A(38)^2*P^2)
 -A(37)*P):Y=
 (A(37)*P^2*A
 (38))
213:Z=1
215:G=Z^3-Z^2-X*
 Z-Y
220:I=3*Z^2-2*Z-
 X
225:A(39)=Z-(G/I
 )
230:N=A(39)^3-A(
 39)^2-X*A(39
 )-Y
235:IF ABS (N)<.
 00001GOTO 25
 0
240:Z=A(39)
241:PAUSE USING
 ;Z
245:GOTO 215
250:V=(A(39)*R*T
 )/P:M=V/(V+A
 (36))
255:RETURN
300:S=R*(LN ((V-
 A(36))/V))-(
 A(35)/(2*A(3
 6)*T^1.5))*
 LN ((V+A(36)
 )/V)+R*LN (V
 /(R*T))
305:S=S*24.2179
310:RETURN
```

Table 6.2
Program 15
Sample Output, Input Data,
Storage Information and Program Notes

```
T1=4.44T2=93.33
P1=13.61P2=170.1
H2=783.4821729--
CAL/GMOL
S2=2.440781515--
CAL/GMOL-K
S1=-5.341132445
Z1=0.967973291V1
=1.620153724H1=-
60.10544917
Z2=9.372564801E-
01V2=1.657096423
E-01H3=-380.7566
773
S3=-11.03615484
H=462.8309448--C
AL/GMOL
S=-3.25424088--C
AL/GMOL-K
```

Input Data: $T_1 = 4.44°C$, $T_2 = 93.33°C$
$P_1 = 13.61$ atm, $P_2 = 170.1$ atm

Storage Information:

A

B   Constants of heat capacity eqn. (5.1)

C

D

R   = gas law constant (0.08206 liter-atm/g mole °K)

K   = (C_p/C_v) ratio

A(27) = T_c, °K

A(28) = P_c, atm

A(29) = 273.16 °K (0 °C)

Program Notes:

1. The physical data for methane were used in the example:
   k = 1.29, A = 4.598, B = 0.01245, C = 2.86E-6 and D = -2.7E-9.

2. Redlich-Kwong EOS was used to evaluate isothermal enthalpy and entropy changes.

Calculation of Total Changes in Enthalpy and Entropy   63

6.2 CALCULATION OF FINAL TEMPERATURE WHEN $\Delta S = 0$

In sizing compressors (or expander) the calculation of the ideal compression, or expansion, process, i.e., $\Delta S = 0$, is frequently the initial calculation made. In the type of calculation the initial conditions of temperature and pressure are known as well as the desired final pressure. However, the final temperature is unknown, but is set by the criterion: $\Delta S = 0$. A trial and error solution if required and equation 6.4 is used.

The solution to this type of problem is given in Program 16 and follows the general procedure of Program 15. The initial conditions, the final pressure and an estimated value of the final temperature are the input data. Alternatively, the initial estimate of the final temperature may be made by applying the ideal gas law to a reversible adiabatic process, or:

$$T_2 = T_1 (P_2/P_1)^{k-1/k} \qquad (6.5)$$

To use this method a value of "zero" is entered as the initial estimate of T_2 and in step 4 the calculations are directed to the solution of equation 6.5 (step 320).

ΔS_1 is calculated, as before, and is a fixed value. The estimated final temperature is used to calculate ΔS_2 and ΔS_3. The summation of these three values is tested against an arbitrarily set criterion--.01 in the program. Specifically, is $abs(\Sigma \Delta S) > .01$? If this condition is not met and depending upon whether ΔS is greater than zero or less than zero, the estimated temperature value is changed by $10°$. T estimated is reduced if the $\Delta S > 0$ and vice versa--steps 65-75 in the program. The Newton technique is then used to make further adjustments in the estimated temperature value:

$$T_3 + T_2 - (\Sigma \Delta S / (\Delta \Sigma \Delta S / \Delta T)) \qquad (6.6)$$

Table 6.3
Program 16
Calculation of Final Temperature for an Isentropic Process

```
1:"A"INPUT A(5          86:A(32)=S              250:Z=A(39):V=(Z
  ),T,A(6),P            90:J=A(32)+A(31             *R*T)/P
2:PRINT "T1=";             )-A(30)              255:S=R*(LN ((V-
  A(5);"P1=";A         95:IF ABS (J)<.              A(36))/V))-(
  (6);"P2=";P             01GOTO 300               A(35)/(2*A(3
3:A(5)=A(5)+A(        100:Q=(J-A(40))/              6)*T^1.5))*
  29)                    10                        LN ((V+A(36)
4:IF T=0GOTO 3        105:A(41)=T-(J/Q              )/V)+R*LN (V
  20                     )                         /(R*T))
5:T=T+A(29)           110:T=A(41)                256:S=S*24.2179
8:A(33)=P:A(34        111:PAUSE USING            260:RETURN
  )=T                    ;T                      280:S=A*LN (T/A(
10:A(35)=(.4278       115:GOSUB 280                  5))+B*(T-A(5
  *R^2*A(27)^2        116:A(31)=S                    ))+(C*(T^2-A
  .5)/A(28):A(        120:GOSUB 200                  (5)^2))/2+(D
  36)=(.0867*R        121:A(32)=S                    *(T^3-A(5)^3
  *A(27)/A(28)        125:A(42)=A(32)+                ))/3
15:P=A(6):T=A(5          A(31)-A(30)             285:RETURN
  )                   130:J=A(42)                300:T=T-A(29)
20:GOSUB 200          135:GOTO 90                305:PRINT "T=";T
25:PRINT "Z1=";       200:A(37)=A(35)/               ;"S=";J;"Z2=
  Z;"V1=";V;"S           (R^2*T^2.5):               ";Z;"V2=";V
  1=";S                  A(38)=A(36)/           307:PRINT "S1=";
26:A(30)=S               (R*T)                      A(30);"S2=";
30:T=A(34)            205:X=(A(38)*P+A                A(31);"S3=";
35:GOSUB 280             (38)^2*P^2-A                A(32)
40:A(31)=S               (37)*P):Y=(A           310:END
45:P=A(33)               (37)*P^2*A(3           320:T=A(5)*(P/A(
50:GOSUB 200             8)):Z=1                    6))^((K-1)/K
52:A(32)=S:A(40       210:G=Z^3-Z^2-X*                )
  )=-A(30)+A(3           Z-Y:I=3*Z^-2           323:PAUSE USING
  1)+A(32)               *Z-X                       ;T
60:IF ABS (A(40       215:A(39)=Z-(G/I           325:GOTO 8
  ))<.01GOTO 3           )
  00                  220:N=A(39)^3-A(
65:IF A(40)>0            39)^2-X*A(39
  ·LET T=T-10            )-Y
70:IF A(40)>0         225:IF ABS (N)<.
  GOTO 35                0001GOTO 250
75:T=T+10             230:Z=A(39)
80:GOSUB 280          231:PAUSE USING
81:A(31)=S               ;Z
85:GOSUB 200          235:GOTO 210
```

Calculation of Total Changes in Enthalpy and Entropy

Table 6.4
Program 16
Storage Information, Input Data,
Storage Information and Program Notes

```
T1=204.4P1=68.04
6P2=6.8046
Z1=9.934394614E-
01V1=5.721346655
E-01S1=-8.569407
276
T=19.1428963S=0.
000549608Z2=9.86
6323027E-01V2=3.
477902163
S1=-8.569407276S
2=-4.692010411S3
=-3.876847257
```

Input Data: $T_1 = 204.4°C$, $T_2 =$ est

$P_1 = 68.04$ atm, $P_2 = 6.8046$ atm

Storage Information: See Table 16.2

Program Notes: See Table 16.2

T_3 is the new estimate of the final temperature. When the abs($\Sigma\Delta S$) > .01, then the final results are printed. Program 16 is given in Table 6-3 and sample output, input data, storage information, and program notes are given in Table 6-4.

To some users a program which determines the final temperature when $\Delta H = 0$ might be useful. The technique used in Program 16 can be used to accomplish this. Subroutines given in Program 15 would be useful in developing this type of program.

7
Correlations for Prediction of Viscosities

Viscosity data are needed in calculations of momentum, heat, and mass transfer rates. Hence, considerable work has been done to develop correlations to predict this transport property. In this chapter, programs are given for the prediction of the viscosities of pure components and binary mixtures of liquids and gases. Non-polar and polar gases are treated separately.

7.1 VISCOSITIES OF PURE LIQUIDS AND BINARIES

Two correlations have been programmed to predict the viscosity of a pure liquid. The first is that of Thomas (1946):

$$\log(8.569\, \eta_L/\rho_L^{.5}) = \theta((1/T_r) - 1) \qquad (7.1)$$

where θ is the property of the structure of the molecule and is calculated from data given by Thomas. In addition, the critical temperature and the liquid density in grams/cm^3 are required. In Program 17, Table 7-1, the latter is calculated by the Spencer-Danner modification of the Rackett equation, Equation 2-1. While the density of the saturated liquid is determined, this value should be sufficiently accurate.

Table 7.1
Program 17
Calculation of the Viscosities of Liquids and Liquid Mixture

```
1:"A"INPUT T
3:PRINT "T=";T
5:T=T+A(29):A(
   31)=T/A(27)
10:V=(R*A(27)/A
   (28))*Z^(1+(
   1-A(31))^(2/
   7))
15:D=M/V
20:N=10^(A(30)*
   (A(31)^-1-1)
   )
25:N=(N*D^.5)/8
   .568
30:PRINT "NU=";
   N;"--CENTIPO
   ISE";"--THOM
   AS"
35:IF A(31)<.76
   GOTO 60
40:IF A(31)>.98
   GOTO 60
45:A(32)=.01517
   4-.02135*A(3
   1)+.0075*A(3
   1)^2
50:A(33)=.04255
   2-.07674*A(3
   1)+.034*A(31
   )^2
53:N=(A(32)+W*A
   (33))/(A(27)
   ^(1/6)/(M^.5
   *A(28)^(2/3)
   ))
55:PRINT "NU=";
   N;"--LETSOU-
   STIEL"
57:END
60:PRINT "CORRE
   LATION-DOES-
   NOT-APPLY"
61:END

70:"B"INPUT X
72:PRINT "X1=";
   X
75:A(42)=N/D:A(
   44)=A(43)/A(
   41)
80:V=(X*M/D)/((
   X*M/D)+((1-X
   )*A(40)/A(41
   ))
85:A(45)=-1.7*
   LN (A(44)/A(
   42))
90:A(46)=.27*LN
   (A(44)/A(42)
   )+(1.3*LN (A
   (44)/A(42)))
   ^.5
95:K=V*A(42)*
   EXP ((1-V)*A
   (46))+(1-V)*
   A(44)*EXP (V
   *A(45))
100:K=K*A(47)
105:PRINT "NU=";
   K;"-CP";"--L
   OBE"
110:END
```

Correlations for Prediction of Viscosities

The second correlation to predict the liquid viscosity is that proposed by Letsou and Stiel (1973):

$$\eta_L \xi = (\eta_L \xi)^0 + \omega (\eta_L \xi)^1 \qquad (7.2)$$

where

$$(\eta_L \xi)^0 = 0.015174 - 0.02135\, T_r + 0.0075\, T_r^2 \qquad (7.3)$$

$$(\eta_L \xi)^1 = 0.042552 - 0.07674\, T_r + 0.0340\, T_r^2 \qquad (7.4)$$

$$\xi = T_c^{1/6} / (M^{1/2}\, P_c^{2/3}) \qquad (7.5)$$

This correlation is restricted to the reduced temperature range $0.76 \leq T_r \leq 0.98$. The lower limit would generally represent a temperature considerably above the normal boiling point. As in the previous correlation only the temperature must be specified.

Program 17 includes programs for both these correlations and in the case of the Letsou and Stiel correlation prints out the message "correlation does not apply" when the reduced temperature does not fall within the required range. The programs are defined as "A".

Program 17 also includes Lobe's correlation (1973) for the prediction of viscosity of a binary liquid mixture:

$$\nu_m = \phi_1 \nu_1 e^{\phi_2 \alpha_2^*} + \phi_2 \nu_2 e^{\phi_1 \alpha_1^*} \qquad (7.6)$$

where

$$\alpha_1^* = -1.7 \ln(\nu_2/\nu_1) \qquad (7.7)$$

$$\alpha_2^* = 0.27 \ln(\nu_2/\nu_1) + (1.3 \ln(\nu_2/\nu_1))^{.5} \qquad (7.8)$$

Table 7.2
Program 17
Sample Output, Input Data, Storage Information
and Program Notes

```
T=4.44                          X1=0.394
NU=1.134635535--                NU=4.65555023-CP
CENTIPOISE--THOM                --LOBE
AS
CORRELATION-DOES
-NOT-APPLY
T=204.4
NU=1.580842671E-
01--CENTIPOISE--
THOMAS
NU=1.510205805E-
01--LETSOU-STIEL
```

"A"   "B"

Input Data: "A" $t = 4.44°C$

"B" $x_1 = 0.394$

Storage Information:

"A"   M   = mole weight

R   = 82.06 gas law constant, (cc-atm/g mole - °K)

w   = acentric factor

Z   = Z_{RA} modified critical compressibility factor

A(27) = T_c, °K

A(28) = ρ_c, atm

A(29) = 273.16°K (0°C)

A(30) = θ

"B"   D   = ρ_1 density, comp 1, gm/cm^3

M   = M_1 mole weight, comp 1

N   = μ_1 viscosity comp 1, cp

A(40) = M_2, mole weight, comp 2

A(41) = ρ_2, density, comp 2, gm/cm^3

A(43) = μ_2, viscosity, comp 2, cp

A(47) = ρ_m, density, solution, gm/cm^3

Program Notes:

1. In "A" the physical data for n-decane have been used. θ = 0.858. Additional physical data are included in the Appendix.

2. In "B" the same example as used by Reid et al. (1977) was used. For the ethyl benzoate (1) and benzyl benzoate (2) system. M_1 = 150.192, ρ_1 = 1.043 gm/cm^3, μ_1 = 2.01 cp, M_2 = 197.936, ρ_2 = 1.112 gm/cm^3, μ_2 = 8.38 cp and ρ_m = 1.091

Correlations for Prediction of Viscosities   71

$$\phi_2 = x_2 V_{L2}/(x_2 V_{L2} + x_1 V_{L1}) \text{ and } \phi_1 = 1 - \phi_2 \qquad (7.9)$$

To use this correlation, the kinematic viscosities as well as molar liquid volumes of the two components and composition are required. This program is defined as "B". Sample output, input data, storage information and program notes for Program 17 are given in Table 7.2.

7.2 GASES

Correlations for the prediction of the viscosities of gases are numerous, since a completely generalized correlation has not been developed. For pure gases, different correlations must be used depending on the nature of the substance; i.e., non-polar or polar. Also, the viscosity of a gas is more pressure dependent than that of a liquid. This effect has also been correlated. Finally, gaseous mixtures present special problems. When dealing with the gas phase, the kinematic viscosity, the ratio of absolute viscosity to the density, is usually employed and the unit is the poises, (grams/cm-sec). Since gas viscosities are small, the micropoise is the unit employed.

7.2.1 Pure Gas-Non-Polar

Two correlations have been programmed for the prediction of the viscosity of a non-polar gas at low pressure. The first is the semi-theoretical correlation of Chapman-Enskog

$$\eta = 26.69((MT)^{.5}/\sigma^2 \Omega_v) \qquad (7.10)$$

where the collision integral, Ω_v, evaluated by Neufeld, Jansen, and Aziz (1972), is:

$$\Omega_v = A/(T*^B) + C/(\exp(DT*)) + E/(\exp(FT*)) \qquad (7.11)$$

and

$$T^* = kT/\varepsilon \tag{7.12}$$

The energy potential parameter, ε, and the molecular diameter, σ, may be calculated by relationships given by Tee, Gotoh and Stewart (1966):

$$\varepsilon/kT_c = 0.7915 + 0.0693\omega \tag{7.13}$$

$$\sigma(P_c/T_c)^{1/3} = 2.3551 - 0.087\omega \tag{7.14}$$

A, B, C, D, E and F are constants.

The second correlation given is the generalized one of Yoon and Thodos (1970):

$$\eta\xi = 4.610 T_r^{0.618} - 2.04\exp(-.449 T_r) + 1.94\exp(-4.058 T_r) + 0.1 \tag{7.15}$$

Program 18 given in Table 7.3 reports the programs for these two correlations. The temperature, $^\circ$C, must be entered to start the calculation and the viscosities, predicted by the equations 7.11 and 7.15 are printed. Sample output, input data, storage information, and program notes are given in Table 7.4. The program is defined as "A".

Jossi, Stiel and Thodos (1962) developed a generalized correlation to predict the effect of pressure on the viscosity of a non-polar gas:

$$[(\eta-\eta^\circ)\xi + 1]^{.25} = 1.0230 + 0.23364\rho_r + 0.58533\rho_r^2 - 0.40758\rho_r^3 + 0.093324\rho_r^4 \tag{7.16}$$

This relationship is valid where $0.1 \leq \rho_r < 3$. To use equation 7.16,

Table 7.3
Program 18
Calculations of Viscosities of Non-Polar Gases

```
1:"A"INPUT T
3:PRINT "T=";T
5:O=(2.3551-.0
  87*W)*(A(27)
  /A(28))^(1/3
  ):E=.7915+.1
  693*W
10:T=T+A(29):A(
  30)=T/(E*A(2
  7)):A(31)=T/
  A(27)
15:Q=(1.16145/(
  A(30)^.14874
  ))+(.5248/
  EXP(.7732*A
  (30)))
20:Q=Q+(2.16178
  /EXP(2.4378
  7*A(30)))
25:N=(26.69*(M*
  T)^.5)/(O^2*
  Q)
30:PRINT "NU=";
  N;"--MPOISES
  ";"--CHAPMAN
  "
35:S=A(27)^(1/6
  )/(M^.5*A(28
  )^(2/3))
40:N=(4.61*A(31
  )^.618-(2.04
  *EXP(-.449*
  A(31))))
45:N=(N+(1.94*
  EXP(-4.058*
  A(31)))+.1)/
  S
50:PRINT "NU=";
  N;"--YOON-TH
  ODOS"
55:END

100:"B"INPUT T,P
103:PRINT "T=";T
  ;"P=";P
105:T=T+A(29):A(
  31)=T/A(27)
110:GOSUB 250
115:IF A(34)<.1
  GOTO 320
120:IF A(34)>3
  GOTO 320
122:S=A(27)^(1/6
  )/(M^.5*A(28
  )^(2/3))
125:J=1.023+.233
  64*A(34)+.58
  533*A(34)^2-
  .40758*A(34)
  ^3+.093324*A
  (34)^4
130:J=J^4:J=(J-1
  )/S:J=J+N
135:PRINT "NU=";
  J;"--MPOISES
  -JOSSI-STIEL
  -THODOS"
140:END
250:A=(.4278/(A(
  28)*A(31)^2.
  5))
252:B=(.0867/(A(
  28)*A(31)))
255:U=(B*P+B^2*P
  ^2-A*P):X=(A
  *P^2*B):Y=1
260:G=Y^3-Y^2-U*
  Y-X:C=3*Y^2-
  2*Y-U
265:A(32)=Y-(G/C
  )
270:F=A(32)^3-A(
  32)^2-U*A(32
  )-X
275:IF ABS(F)<.
  0001GOTO 300
280:Y=A(32)
285:GOTO 260

300:Y=A(32):V=(Y
  *R*T)/P:A(33
  )=(Z*R*A(27)
  )/A(28)
305:PRINT "V=";V
  ;"--L/GMOL--
  R-K"
310:A(34)=A(33)/
  V
315:RETURN
320:PRINT "CORRE
  LATION-DOES-
  NOT-APPLY"
325:END
```

Table 7.4
Program 18
Sample Output, Input Data,
Storage Information, and Program Notes

```
     T=171.11                  T=26.67P=27.218
     NU=133.9397179--          V=6.843335622E-0
     MPOISES--CHAPMAN          1--L/GMOL--R-K
     NU=135.8802943--          NU=106.3983695--
     YOON-THODOS               MPOISES-JOSSI-ST
                               IEL-THODOS
```

"A" "B"

Input Data: "A" t = 171.11°C

"B" 5 = 26.67°C, P = 27.218 atm

Storage Information:

M   = mole weight

N   = $\eta°$ viscosity at low pressure, mp

R   = gas law constant (.08206 liter-atm/g mole °K)

w   = acentric factor

Z   = Z_c, critical compressibility factor

A(27) = T_c, °K

A(28) = P_c, atm

A(29) = 273.16°K (0°C)

Program Notes:
1. $\eta°$, R and Z_c are the additional data required for "B".
2. Physical data for ethane were used in the examples and $\eta° = 93.84$ μp.
3. The value of $\eta°$ could have been calculated by "A" and used in "B".

---

the reduced density as well as the viscosity at low pressure, $\eta°$, must be known. The latter may be estimated by either of the correlations discussed above. The Redlich-Kwong EOS, Equation 1.1, has been included as a subroutine to calculate the density, which then permits the determination of the reduced density. A test is also made to ascertain whether the reduced density falls within the required range of values. If it does not, the statement "Correlation Does Not Apply" is printed.

Correlations for Prediction of Viscosities

The program is included in Program 18 and is defined as "B". The temperature and pressure are entered to start to calculations and the volume calculated by the Redlich-Kwong EOS and the viscosity in micropoises are calculated and printed. Sample output and other pertinent information are given in Table 7.4.

7.2.2 Pure Gases - Polar

At low pressure the viscosity of a polar gas may be predicted by a modified form of the Chapman-Enskog relationship--equation 7.11. The collision integral, Ω_v, is modified by:

$$\Omega_v(\text{polar}) = \Omega_v(\text{non-polar}) + 0.2 \, \delta^2/T^* \quad (7.17)$$

For polar gases the terms σ and ε/k as well as δ are difficult to calculate and usually are obtained from tabulations. This correlation is programmed in Program 19 and defined as "A" and is shown in Table 7.5. Sample output and other information are given in Table 7.6.

Also, in Program 19 is the correlation of Stiel and Thodos (1964) to determine the viscosity of a polar gas at an elevated pressure. Similarly to the case of non-polar gases, the effect of pressure on viscosity is related to the reduced density, but in this case a series of relationships are required as follows:

if $\rho_r \leq 0.1$

$$(\eta - \eta^\circ)\xi = 1.656 \rho_r^{1.111} \quad (7.18)$$

if $0.1 \leq \rho_r \leq 0.9$

$$(\eta - \eta^\circ)\xi = 0.0607(9.045 \rho_r + 0.63)^{1.739} \quad (7.19)$$

if $0.9 \leq \rho_r \leq 2.6$

$$\log(4 - \log((\eta - \eta^\circ)\xi)) = 0.6439 - 0.1005 \rho_r - \Delta \quad (7\text{-}20)$$

Table 7.5
Program 19
Calculation of Viscosities of Polar Gases

```
1:"A"INPUT T
3:PRINT "T=";T
5:T=T+A(29):A(
   30)=T/E:A(31
   )=T/A(27)
10:Q=(1.16145/(
   A(30)^.14874
   ))+(.5248/
   EXP (.7732*A
   (30)))
15:Q=Q+(2.16178
   /EXP (2.4378
   7*A(30)))
20:Q=Q+((.2*D^2
   )/A(30))
25:N=(26.69*(M*
   T)^.5/(O^2*Q
   )
30:PRINT "NU=";
   N;"--MPOISES
   --MOD-CHAPMA
   N"
35:END

100:"B"INPUT T,P
103:PRINT "T=";T
    ;"P=";P
105:T=T+A(29):A(
    31)=T/A(27)
107:S=A(27)^(1/6
    )/(M^.5*A(28
    )^(2/3))
110:GOSUB 250
115:IF A(34)<.1
    GOTO 130
116:IF A(34)<.9
    GOTO 135
117:IF A(34)<2.2
    GOTO 140
118:IF A(34)<2.6
    GOTO 145
120:IF A(34)>2.6
    GOTO 320
130:J=(1.656*A(3
    4)^1.111)/S
131:J=N+J
132:GOTO 155
135:J=(.0607*(9.
    045*A(34)+.6
    3)^1.739)/S
136:J=N+J
137:GOTO 155
140:J=10^(.6439-
    .1005*A(34))
141:J=J-4:J=-J:J
    =10^J/S
142:J=N+J
143:GOTO 155
145:J=10^(.6439-
    .1005*A(34)-
    (.000475*(A(
    34)^3-10.65)
    ^2))
146:J=J-4:J=-J:J
    =(10^J)/S
148:J=N+J
155:PRINT "NU=";
    J;"--MPOISES
    --STIEL-THOD
    OS"
156:END

250:A=(.4278/(A(
    28)*A(31)^2.
    5))
252:B=(.0867/(A(
    28)*A(31)))
255:U=(B*P+B^2*P
    ^2-A*P):X=(A
    *P^2+B):Y=1
260:G=Y^3-Y^2-U*
    Y-X:C=3*Y^2-
    2*Y-U
265:A(32)=Y-(G/C
    )
270:F=A(32)^3-A(
    32)^2-U*A(32
    )-X
275:IF ABS (F)<.
    0001GOTO 300
280:Y=A(32)
285:GOTO 260
300:Y=A(32):V=(Y
    *R*T)/P:A(33
    )=(Z*R*A(27)
    )/A(28)
305:PRINT "V=";V
    ;"--L/GMOL--
    R-K"
310:A(34)=A(33)/
    V
315:RETURN
320:PRINT "CORRE
    LATION-DOES-
    NOT-APPLY"
325:END
```

Correlations for Prediction of Viscosities

Table 7.6
Program 19
Sample Output, Input Data,
Storage Information and Program Notes

```
                              T=171.11P=68.046
    T=171.11                  V=4.412688889E-0
    NU=154.9874395--          1--L/GMOL--R-K
    MPOISES--MOD-CHA          NU=162.8146628--
    PMAN                      MPOISES--STIEL-T
                              HODOS
```

                 "A"   "B"

Input Data: "A" $171.11^\circ C$

                "B" $171.11^\circ C$ and 68.047 atm

Storage Information:

    D   = δ, polar parameter

    E   = ε/kT_c (ε, energy-potential parameter)

    M   = molecular weight

    O   = σ, molecular diameter, A°

    R   = gas law constant (.08206 liter-atm/g $mole^\circ K$)

    W   = acentric factor

    Z   = Z_c critical compressibility factor

    A(27) = T_c, $^\circ K$

    A(28) = P_c, atm

    A(29) = $273.16^\circ K$, ($0^\circ C$)

Program Notes:

1. Physical constants for ammonia were used throughout.
   $\delta = 0.7$, $\varepsilon/kT_c = 358$ and $\sigma = 3.15$.
2. Value of η° used in "B" was calculated in "A".

---

if $0.9 \leq \rho_r < 2.2$ $\Delta = 0$

if $2.2 < \rho_r < 2.6$

$$\Delta = 0.000475 \, (\rho_r^3 - 10.65)^2 \qquad (7.21)$$

If the reduced density is greater than 2.6, the correlation does not apply.

Equations 7.18 through 7.11 have been programmed and based on the reduced density value, the appropriate relationship is selected. The Redlich-Kwong EOS is included in the program to calculate the density. This program is designated as "B" and requires temperature and pressure values as input data. Sample output and other pertinent information are given in Table 7.6.

7.3 BINARY GASEOUS MIXTURES

Under low pressure, the viscosity of a binary gaseous mixture may be determined from the correlation of Wilke (1950):

$$\eta_m = \sum_{i=1}^{\eta} (y_i \eta_i / \sum_{j=1}^{\eta} y_i \phi_i) \qquad (7.22)$$

where, for a binary mixture,

$$\phi_{12} = [1 + (\eta_1/\eta_2)^{.5} (M_2/M_1)^{.25}]^2 / [8(1 + (M_1/M_2))]^{.5} \qquad (7.23)$$

and

$$\phi_{21} = (\eta_2/\eta_1)(M_1/M_2) \phi_{12} \qquad (7.24)$$

Combining 7.23 and 7.24, 7.22 yields:

$$\eta_m = (y_1 \eta_1)/(y_1 + y_2 \phi_{12}) + (y_2 \eta_2)/(y_2 + y_1 \phi_{21}) \qquad (7.25)$$

The program for relationship is given in Program 20 included in Table 7.7. Sample output, input data, and other information is in Table 7.8.

For mixture of non-polar gases under pressure, Dean and Stiel (1965) correlated the viscosity difference $(\eta_m - \eta_m^o)$ using the generalized relationship:

Correlations for Prediction of Viscosities

$$(\eta_m - \eta_m^o)\xi_m = 1.08[\exp(1.439\rho_{rm}) - \exp(-1.111\rho_{rm}^{1.858})] \quad (7.26)$$

where the critical properties after Prausnitz and Gunn (1958) are defined as:

$$T_{cm} = \sum_i y_i T_{ci} \quad (7.27)$$

$$Z_{cm} = \sum_i y_i Z_{ci} \quad (7.28)$$

$$V_{cm} = \sum_i y_i V_{ci} \quad (7.29)$$

Table 7.7
Program 20
Calculation of Viscosities of Gaseous Mixtures

```
 1:"A"INPUT Y
 2:PRINT "Y1=";
   Y
 3:X=1-Y
10:H=(1+((N/A(3
   5))^.5*(A(34
   )/M)^.25))^2
   /(8*(1+(M/A(
   34))))^.5
15:A(36)=(A(35)
   /N)*(M/A(34)
   )*H
20:A(37)=((Y*N)
   /(Y+X*H))+((
   X*A(35))/(X+
   Y*A(36)))
25:PRINT "VIS="
   ;A(37);"--MP
   OISES--WILKE
   "
30:END

35:"B"INPUT Y
38:PRINT "Y1=";
   Y
41:X=1-Y
45:V=(R*Z*A(27)
   )/A(28):A(43
   )=(A(32)*R*A
   (30))/A(31)
50:A(38)=Y*A(27
   )+X*A(30):A(
   40)=Y*V+X*A(
   43):A(39)=Y*
   Z+X*A(32)
55:A(41)=(A(39)
   *R*A(38))/A(
   40)
60:A(44)=Y*M+X*
   A(34)
65:S=A(38)^(1/6
   )/(A(44)^.5*
   A(41)^(2/3))
70:A(45)=A(40)/
   A(33)
75:J=(1.08*(EXP
   (1.439*A(45)
   )-EXP (-1.11
   1*A(45)^1.85
   8)))/S
80:J=J+A(37)
85:PRINT "VIS="
   ;J;"--MPOISE
   S--DEAN-STIE
   L"
90:END
```

80   BASIC Programs for Chemical Engineering Design

Table 7.8
Program 20
Sample Output, Input Data
Storage Information, and Program Notes

```
Y1=0.384
VIS=86.22754905-
-MPOISES--WILKE
Y1=0.384
VIS=425.5242444-
-MPOISES--DEAN-S
TIEL
```

Input Data: $y_1 = 0.384$ mole fraction methane (1) (with n-butane)

Storage Information:

$M \quad = M_1^*$

$N \quad = \eta_1^*$, μp

$R \quad$ = gas law constant (.08206 liter-atm/g mole-°K)

$Z \quad = Z_{c1}$

$A(27) = T_{c1}$, °K

$A(28) = P_{c1}$, atm

$A(30) = T_{c2}$, °K

$A(31) = P_{c2}$, atm

$A(32) = Z_{c2}$

$A(33) = V_m^{**}$, l/g mole

$A(34) = M_2^{**}$

$A(35) = \eta_2^*$, μp

$A(37) = \eta_m^{o**}$, μp

* Required for solution of equation 7.25 ("A")

** Required for solution of equation 7.26 ("B")

Program Notes:

1. The methane (1) - n-butane (2) system was used in both examples.
2. In "A", t = 37.78°C, for methane η = 116.1μp and for n-butane η = 77.09μp.
3. In "B", t = 237.8°C and P = 340.2 atm, η_m^o = 162μP, V_m = .130 liters/g mole

Correlations for Prediction of Viscosities

$$P_{cm} = (Z_{cm}RT_{cm})/V_{cm} \tag{7.30}$$

From these four equations ξ_m and ρ_{cm} are calculated and, provided a value of V_m is available, ρ_{rm} can be determined. If the user desires, an EOS could be included in this program to calculate V_m. Also, the value of η_m^o calculated by equation 7.25 may be used, or the user may desire to enter a value. Programs for the solution of equations 7.25 and 7.26 are given in Program 20 (Table 7.7). Sample output, input data, storage information and program notes are given in Table 7.8.

8
Correlations for Prediction of Thermal Conductivity

The transport property thermal conductivity is necessary in heat transfer calculation and, as such, its determination has been the subject of many investigations. Numerous correlations have been proposed. In the following sections, programs are given for the prediction of thermal conductivity values for pure liquid and mixture, for pure gases at both low and high pressure, and for gaseous mixtures.

8.1 PURE LIQUIDS AND MIXTURES

Sato (see Reid et al., 1977) proposed the relationship

$$k_{Lb} = 2.64 \times 10^{-3}/M^{.5} \tag{8.1}$$

for the prediction of the thermal conductivity of a liquid at its normal boiling point and Reidel (1949, 1951) suggested the general relationship:

$$k_L = A(1 + (20/3)(1 - T_r)^{2/3}) \qquad (8.2)$$

The constant A can be eliminated by combining these two equations into:

$$k_L = ((2.64 \times 10^{-3})/M^{.5}) \cdot ((3 + 20(1 - T_r)^{2/3} / (3 + 20(1 - T_{br})^{2/3})) \qquad (8.3)$$

Thus, k values may be predicted from very limited data; critical temperature, normal boiling temperature, and molecular weight.

A second correlation based on Missenard's relationship (1965):

$$k_{Lo} = (84 \times 10^{-6} (T_b \rho_o)^{.5} Cp_{Lo})/(M^{.5} N'^{.25}) \qquad (8.4)$$

Combining equations 8.4 and 8.2 gives:

$$k_L = k_{Lo}((3 + 20(1 - T_r)^{2/3})/(3 + 20(1 - 273/T_c))^{2/3})) \qquad (8.5)$$

To use equation 8.5, the liquid density and heat capacity are required in addition to the information required to solve Equation 8.3. N is simply the number of atoms in the molecule of the substance.

The Spencer and Danner modification of the Rackett equation, Equation 2.1, may be used to calculate the liquid density at 0°C and the Bondi modification (1966) of the Rowlinson correlation (1969):

$$(C_{pL} - C_p^*)/R = 2.56 + 0.436(1 - T_r)^{-1} + \omega(2.91 + 4.28(1 - T_r)^{1/3} T_r^{-1} + 0.296(1 - T_r)^{-1}) \qquad (8.6)$$

Correlations for Prediction of Thermal Conductivity 85

may be used to calculate C_L at $0°C$. The ideal gas phase heat capacity can be calculated from Equation 5.1.

The two correlations, Equation 8.3 and 8.5, are included in Program 21 and defined as "A". Other than the stored information, the only input data is the temperature. The temperature and thermal conductivity values are printed out. Program 21 is included in Table 8.1 and Table 8.2 shows a sample output and gives the other information necessary to use the programs.

For liquid mixtures, Li (1976) has suggested:

$$k_m = \Sigma\Sigma \phi_i \phi_j k_{ij} \tag{8.7}$$

where

$$k_{ij} = 2(k_i^{-1} + k_j^{-1})^{-1} \tag{8.8}$$

and

$$\phi_i = x_i V_i / \Sigma x_j V_j \tag{8.9}$$

and Jordan (1961) has proposed:

$$k_m/(k_1^{w_1} k_2^{w_2}) = \exp(|k_2 - k_1|) - .5(k_2 + k_1))^{(w_1 w_2)} \tag{8.10}$$

The w's represent weight fractions. The solution of Equations 8.7 through 8.10 are included in Program 21, Table 8.1. Sample output, input data and other information are given in Table 8.2. The program for the determination of thermal conductivities of mixture are defined as "B" and require that the mole fraction of component 1 be entered to start the calculations.

Table 8.1
Program 21
Programs for the Thermal Conductivities of Liquids

```
1:"A"INPUT T
3:PRINT "T=";T
5:T=T+A(29):A(
  31)=T/A(27)
6:A(32)=A(30)/
  A(27)
10:K=(.00264/M^
  .5)*((3+(20*
  (1-A(31))^(2
  /3)))/(3+(20
  *(1-A(32))^(
  2/3))))
15:PRINT "K=";K
  ;"-CAL/CM-S-
  K--SATO"
20:V=(82.06*A(2
  7)/A(28))*Z^
  (1+(1-(A(29)
  /A(27)))^(2/
  7))
25:E=A+B*A(29)+
  C*A(29)^2+D*
  A(29)^3
30:A(33)=A(29)/
  A(27)
35:F=2.56+.436*
  (1-A(33))^-1
40:F=F+(W*(2.91
  +(4.28*(1-A(
  33))^(1/3)*A
  (33)^-1+.296
  *(1-A(33))^-
  1
45:G=F*R+E
50:K=(84*10^-6*
  (A(30)/V)^.5
  *G)/(M^.5*N^
  .25)
55:K=K*((3+(20*
  (1-A(31))^(2
  /3)))/(3+(20
  *(1-(273/A(2
  7))^(2/3)))
  )
60:PRINT "K=";K
  ;"--MISSENAR
  D"
65:END
70:"B"INPUT Y
72:PRINT "Y1=";
  Y
75:X=1-Y
80:P=Y*A(35)/(Y
  *A(35)+X*A(3
  6)):Q=1-P
85:O=P^2*K+(4*P
  *Q*(K^-1+A(3
  4)^-1)^-1)+Q
  ^2*A(34)
90:PRINT "K=";O
  ;"--CAL/CM-S
  -K--LI"
95:W=Y*M/(Y*M+X
  *A(37)):A(38
  )=1-W
100:O=(EXP (ABS
  (A(34)-K))-(
  .5*(A(34)+K)
  ))^(W*A(38))
105:O=O*(K^W*A(3
  4)^A(38))
110:PRINT "K=";O
  ;"--JORDAN"
115:END
```

Table 8.2
Program 21
Sample Output, Input Data
Storage Information and Program Notes

```
T=4.44                          Y1=0.2
K=3.405295919E-0                K=5.5357691E-04-
4-CAL/CM-S-K--SA                -CAL/CM-S-K--LI
TO                              K=5.654022933E-0
K=3.0741852E-04-                4--JORDAN
-MISSENARD
```

                      "A"   "B"

Input Data: "A" $t = 4.44\,^\circ C$

             "B" $y_1 = 0.2$ mole fraction of

Storage Information "A":

    A

    B

           constants of heat capacity expression

    C

    D

    M   = molecular weight

    N   = number of atoms in molecule

    R   = gas law constant (1.987 cal/g mole - $^\circ K$)

    w   = acentric factor

    Z   = Z_{RA}

    A(27) = T_c, $^\circ K$

    A(28) = P_c, atm

    A(29) = $273.16\,^\circ K$ ($0\,^\circ C$)

    A(30) = T_b, $^\circ K$

Storage Information "B":

    K   = k_1, cal/cm -sec-$^\circ K$, comp 1

    M   = M_1, molecular weight, comp 1

    A(34) = k_2, cal/cm-sec-$^\circ K$, comp 1

    A(35) = V_1, volume, cm^3/g mol, comp 1

    A(36) = V_2, volume, cm^3/g mol, comp 2

    A(37) = M_2, molecular weight, comp 2

Program Notes:

1. The physical properties of n-butane were used in "A". N = 14, and A = 2.266, B = .07913, C = -2.647×10^{-5}, D = $-.674 \times 10^{-9}$.

2. In "B" water (1) - methanol (2) - $0\,^\circ C$. For water k = 1.352×10^{-3}, V = 18.0, M = 18.015. For methanol k = 5.002×10^{-4}, V = 37.8, M = 32.042.

8.2 Pure Gases - Low and High Pressure

Thermal conductivities of a pure gas at low pressure may be predicted by relationship proposed by Eucken (see Liley, 1959) and by Stiel and Thodos (1964). These are:

$$kM/\eta = C_v + 4.47 \quad \text{(Eucken)} \tag{8.11}$$

$$kM/\eta = 1.32 C_v + 3.52 \quad \text{(Modified Eucken)} \tag{8.12}$$

$$kM/\eta = 1.15 C_v + 4.04 \quad \text{(Stiel-Thodos)} \tag{8.13}$$

To use these equations, C_v is determined by:

$$C_v^* = C_p^* - R \tag{8.14}$$

and C_p^* by Equation 5.1. The viscosity can be determined by the correlation of Yoon and Thodos, Equation 7.15.

To calculate the thermal conductivity of a gas at high pressure Stiel and Thodos (1964) proposed a series of relationships which permit the evaluated difference of the thermal conductivities at high and low pressure.

These are

$$(k - k^o) \Gamma z_c^5 = (14.0 \times 10^{-8})(e^{0.53\rho_r} - 1) \tag{8.15}$$

when $\rho_r < .5$

$$(k - k^o) \Gamma z_c^5 = (13.1 \times 10^{-8})(e^{0.67\rho_r} - 1.069) \tag{8.16}$$

Table 8.3
Program 22
Programs for the Thermal Conductivities of Pure Gases

```
 1:"A"INPUT T
 3:PRINT "T=";T
 5:T=T+A(29):A(
   30)=T/A(27)
10:S=A(27)^(1/6
   )/(M^.5*A(28
   )^(2/3))
15:N=(4.61*A(30
   )^.618-2.04*
   EXP (-.449*A
   (30))+1.94*
   EXP (-4.058*
   A(30))+.1)/S
20:F=A+B*T+C*T^
   2+D*T^3:E=F-
   R
25:K=(N*(E+4.47
   ))/M
26:K=K*10^-6
27:PRINT "K=";K
   ;"--CAL/CM-S
   -K--EUCKEN"
30:K=(N*(1.32*E
   +3.52))/M
31:K=K*10^-6
33:PRINT "K=";K
   ;"--MOD-EUCK
   EN"
35:K=(N*(1.15*E
   +4.04))/M
36:K=K*10^-6
37:PRINT "K=";K
   ;"--STIEL-TH
   ODOS"
40:END

45:"B"INPUT T,P
47:PRINT "T=";T
   ;"P=";P
50:T=T+A(29):G=
   (A(27)^(1/6)
   *M^.5)/A(28)
   ^(2/3)
52:A(30)=T/A(27
   )
55:A(40)=.4278/
   (A(28)*A(30)
   ^2.5):A(41)=
   .0867/(A(28)
   *A(30))
60:A(42)=(A(41)
   *P+A(41)^2*P
   ^2-A(40)*P):
   Z=1
65:A(43)=(A(40)
   *P^2*A(41))
70:J=Z^3-Z^2-A(
   42)*Z-A(43):
   I=3*Z^2-2*Z-
   A(42)
75:X=Z-(J/I)
80:Q=X^3-X^2-A(
   42)*X-A(43)
85:IF ABS (Q)<.
   00001GOTO 10
   0
90:Z=X
91:PAUSE USING
   ;Z
95:GOTO 70
100:Z=X:V=(Z*.08
   206*T)/P
105:A(32)=(A(31)
   *.08206*A(27
   ))/A(28):A(3
   3)=A(32)/V
110:PRINT "Z=";Z
   ;"V=";V

115:PAUSE USING
   ;A(33)
120:IF A(33)<.5
   GOTO 130
121:IF A(33)<2
   GOTO 135
122:IF A(33)<2.8
   GOTO 140
123:IF A(33)>2.8
   GOTO 145
130:A(34)=(14*10
   ^-8*(EXP (.5
   35*A(33))-1)
   )/(G*A(31)^5
   )
132:L=K+A(34)
133:PRINT "K=";L
   ;"--CAL/CM-S
   -K--STIEL-TH
   ODOS"
134:END
135:A(34)=(13.1*
   10^-8*(EXP (
   .67*A(33))-1
   .069))/(G*A(
   31)^5)
137:GOTO 132
140:A(34)=(2.976
   *10^-8*(EXP
   (1.155*A(33)
   )+2.016))/(G
   *A(31)^5)
142:GOTO 132
145:PRINT "CORRE
   LATION-DOES-
   NOT-APPLY"
146:END
```

Table 8.4
Program 22
Sample Output, Input Data,
Storage Information and Program Notes

T=37.73
K=7.912430415E-05--CAL/CM-S-K--EUCKEN
K=8.754719609E-05--MOD-EUCKEN
K=8.318122806E-05--STIEL-THODOS

T=37.73 P=68.05
Z=9.048737021E-01
1V=3.392330075E-01
K=9.903286707E-05--CAL/CM-S-K--STIEL-THODOS

"A"

T=37.73 P=340.2
Z=9.921875006E-01
1V=7.440424391E-02
K=2.035476086E-04--CAL/CM-S-K--STIEL-THODOS

"B"

Input Data: "A" t = 37.73°C

"B" t = 37.37°C and P = 68.05 atm

Storage Information:

  A
  B
  C   constant of heat capacity expression
  D
  M   = molecular weight
  R   = gas law constant (1.987 cal/g mole - °K)
  A(27) = T_c, °K
  A(28) = P_c, atm
  A(29) = 273.16°K (0°C)
  A(31) = Z_c, critical compressibility factor (for "B" only)

Program Notes:

1. Physical properties of methane used in the example.
2. In "A" C_p^* is calculated by equation 5.1 and is labelled "F". C_v is labelled "E" and is calculated by equation 7.44--step 20. The user may enter a value of C_v and eliminate this step.
3. In "B" the density of the gas is calculated by the Redlich-Kwong EOS. If the user desires to enter a density value, this may be done and steps 40-100, 110, and 115 may be eliminated. The critical volume is calculated from T_c, P_c, and Z_c.

Correlations for Prediction of Thermal Conductivity

when $0.5 < \rho_r < 2.0$

$$(k-k^o)\Gamma Z_c^5 = (2.967 \times 10^{-8})(e^{1.155\rho_r} + 2.016) \quad (8.17)$$

when $2.0 < \rho_r < 2.8$

If the value of the reduced density is greater than 2.8, then the correlation does not apply and this message is printed.

Program 21, Table 8.3, gives the programs for the calculation of thermal conductivities of pure gases at either low or high pressure. The former is defined as "A" and the latter as "B".

Included in the "B" program is a subroutine for the calculations of the density of the gas by the Redlich-Kwong EOS, equation 1.1. With this datum and a knowledge of the critical properties of the substance, the reduced density is determined. The proper equation 8.5, 8.6 or 8.7 is then selected and the calculations made. The user can substitute another equation of state, or simply enter the density if it is known. These modifications are simple to make.

Sample output is shown in Table 8.4 as well as input data, storage information and program notes.

8.3 BINARY GASEOUS MIXTURES

For binary gaseous mixtures, Wassiljewa (1904) recommended the expression:

$$k_m = (y_1 k_1/(y_1 + A_{12} y_2)) + (y_2 k_2/(y_2 + y_1 A_{21})) \quad (8.18)$$

For the interaction term, A_{12} for example, Lindsay and Bromley (1950) proposed:

$$A_{12} = 1/4[1 + ((\eta_1/\eta_2)(M_2/M_1)^{.75}(T + S_1/T + S_2))^{.5}]^2$$

$$(T + S_{12}/T + S_1) \quad (8.19)$$

where

$$S_1 = 1.5T_{bl} \qquad (8.20)$$

$$S_{12} = S_{21} = C_s(S_1 S_2)^{1/2} \qquad (8.21)$$

and C_s is usually equal to unity. A_{21} is obtained by equation 8.19 by simply interchanging the subscripts, 1 and 2. Program 22, shown in Table 8.5, gives the program for equations 8.18 to 8.21.

In Program 23 the thermal conductivities, viscosities, and normal boiling points were summed and values for these items are

Table 8.5
Program 23
Program for the Thermal Conductivity of Gaseous Mixtures

```
1:"A"INPUT T,Y
3:PRINT "T=";T
 ;"Y1=";Y
5:T=T+A(29):X=
  1-Y:S=1.5*A(
  30):A(36)=1.
  5*A(33):A(37
  )=(S*A(36))^
  .5
10:A=.25*(1+((N
   /A(35))*(A(3
   4)/M)^.75*((
   T+S)/(T+A(36
   )))^.5)^2
12:A=A*((T+A(37
   ))/(T+S))
15:B=.25*(1+((A
   (35)/N)*(M/A
   (34))^.75*((
   T+A(36))/(T+
   S)))^.5)^2
17:B=B*((T+A(37
   ))/(T+A(36))
   )
20:A(39)=((Y*K)
   /((Y+A*X))+(
   (X*A(38))/(X
   +Y*B))
25:PRINT "K=";A
   (39);"--CAL/
   CM-S-K--WASS
   ILJEWA"
30:END
```

Correlations for Prediction of Thermal Conductivity   93

Table 8.6
Program 23
Sample Output, Input Data,
Storage Information, Program Notes

```
T=4.44Y1=0.394
K=4.217956646E-0
5--CAL/CM-S-K--W
ASSILJEWA
```

Input Data: $t = 4.44°C$

$y_1 = 0.394$

Storage Information:

K = k_1 cal/cm^2 - s - °K
M = molecular weight, comp 1
N = viscosity, comp 1, mpoises
$A(27)$ = T_{c1}, °K
$A(28)$ = P_{c1}, atm
$A(29)$ = 273.16°K (0°C)
$A(30)$ = T_{b1}, °K
$A(31)$ = T_{c2}, °K
$A(32)$ = P_{c2}, atm
$A(33)$ = T_{c2}, °K
$A(34)$ = molecular weight, comp 2
$A(35)$ = viscosity, comp 2, mpoise
$A(38)$ = k_2 cal/cm^2 - s - °K

Program Notes:
1. For the example, methane is comp 1 and n-butane comp 2. For methane $k = 72.126 \times 10^6$ cal/cm-sec-°K and $\nu = 103.3$ μp For n-butane $k = 32.98 \times 10^6$ and $\nu = 70.32$.
2. A_{12} is labelled A and A_{21}, B, and y_2 is x.

stored before the program is used. However, the user may wish to include, as subroutines, programs to calculate all or part of this information. For example, the pure component thermal conductivities could be calculated from programs given immediately above.

Sample output, input data, and other information is given in Table 8.6.

9
Correlations for Prediction of Diffusion Coefficients

The third transport property is the diffusivity, or diffusion coefficient. This property is important in mass transfer operations and as such, appears in the correlations of mass transfer coefficients. In the following sections programs are given for the calculations of values of diffusion coefficients for liquids and gases. In the latter case, mixtures containing polar compounds must be treated differently than those containing only non-polar substances.

9.1 DIFFUSION COEFFICIENTS IN THE LIQUID PHASE

The value of the diffusion coefficient in liquid phase is dependent upon the composition of the liquid mixture. Since the mechanism of diffusion is very complex, the available correlations apply to dilute solutions only. That is the concentration of the solute - the diffusing component - in the solvent is low, usually less than 10%. Within this restriction, composition is not too important and is disregarded. However, it follows that the "solute" and the "solvent" must be specified.

Three correlations to predict values of diffusion coefficients in the liquid phase are in Program 25, Table 9.1, and defined as "A". The first is that of Wilke and Chang (1955):

$$D_{12} = 7.4 \times 10^{-8} (\phi M_2)^{.5} T/\eta_2 V_1^{0.6} \tag{9.1}$$

where the sbuscript "1" refers to the solute and "2" to the solvent and ϕ is the association factor of solvent - a dimensionless quantity.

The next correlation program used is Scheibel's (1954):

$$D_{12} = KT/\eta_2 V_1^{1/3} \tag{9.2}$$

where

$$K = 8.2 \times 10^{-8} (1 + (3V_2/V_1)^{2/3}) \tag{9.3}$$

Exceptions to equation 9.3 are explained in the program notes in Table 9.2. The third correlation programmed is that of Reddy and Doraiswang (1967):

$$D_{12} = K' M_2^{0.5} T/\eta_2 (V_1 V_2)^{1/3} \tag{9.4}$$

where

$$K' = 10 \times 10^{-8} \text{ if } V_2/V_1 \leq 1.5$$

and 8.5×10^{-8} if the volumetric ratio is ≥ 1.5.

To use the programs for equations 9.1 through 9.4, the temperature is entered. In addition volumetric and viscosity data are required. The former are calculated by the modified Rackett,

Prediction of Diffusion Coefficients

Table 9.1
Program 24
Programs for Liquid and Non-polar
Gas Phase Diffusion Coefficients

```
1:"A"INPUT T
3:PRINT "T=";T
5:T=T+A(29)
8:V=(82.06*A(2
  7)/A(28))*Z^
  (1+(1-(A(30)
  /A(27)))^(2/
  7))
10:A(37)=(82.06
  *A(31)/A(32)
  )*A(34)^(1+(
  1-(A(33)/A(3
  1)))^(2/7))
15:D=7.4*10^-8*
  (((P*A(35))^
  .5*T)/((A(36
  )*V^.6))
18:PRINT "D12="
  ;D;"--CM^2/S
  EC--WILKE-CH
  ANG"
20:IF V<2*A(37)
  GOTO 34
25:K=8.2*10^-8*
  (1+(3*A(37)/
  V)^(2/3))
27:GOTO 35
34:K=18.9*10^-8
35:D=((K*T)/(A(
  36)*V^(1/3))
  )
40:PRINT "D12="
  ;D;"--SCHEIB
  EL"
45:L=A(37)/V
50:IF L<1.5GOTO
  59
52:K=8.5*10^-8
55:GOTO 60
59:K=10*10^-8
60:D=(K*A(35)^.
  5*T)/(A(36)*
  (V*A(37))^(1
  /3))
65:PRINT "D12="
  ;D;"--REDDY-
  DORAISWAMY"
70:END

75:"B"INPUT T,P
77:PRINT "T=";T
  ;"P=";P
80:T=T+A(29)
85:O=(2.3551-(.
  087*W))/(A(2
  8)/A(27))^(1
  /3)
87:Q=(2.3551-(.
  087*A(33)))/
  (A(32)/A(31)
  )^(1/3)
90:O=(O+Q)/2
95:E=(.7915+(.1
  693*W))*A(27
  ):F=(.7915+(
  .1693*A(33))
  )*A(31)
97:E=(E*F)^.5
100:A(36)=T/E
105:I=(1.06036/(
  A(36)^.1561)
  )+(.193/EXP
  (.47635*A(36
  )))
107:I=I+(1.03587
  /EXP (1.5299
  6*A(36)))+(1
  .76474/EXP (
  3.89411*A(36
  )))
110:D=(.001858*T
  ^(3/2)*((M+A
  (35))/(M*A(3
  5)))^.5/(P*O
  ^2*I)
115:PRINT "D12="
  ;D;"--CM^2/S
  EC--CHAPMAN-
  COWLING"
120:D=(D/.001858
  )*(.00217-(.
  0005*((M+A(3
  5))/(M*A(35)
  ))^.5))
125:PRINT "D12="
  ;D;"--WILKE-
  LEE-MOD"
130:END
```

Table 9.2
Program 24
Sample Output, Input Data,
Storage Information, and Program Notes

T=40.
D12=2.39290381E-05--CM^2/SEC--WILKE-CHANG
D12=2.461001168E-05--SCHEIBEL
D12=2.512302306E-05--REDDY-DORAISWAMY

T=104.4P=1.
D12=8.013726804E-02--CM^2/SEC--CHAPMAN-COWLING
D12=8.902653306E-02--WILKE-LEE-MOD

"A"   "B"

Input Data: "A"   t = 40°C
              "B"   t = 104.4°C, P = 1 atm

Storage Information:

    P   = ϕ*, association parameter for the solvent

    M   = molecular weight - comp 1

    w   = acentric factor - comp 1**

    Z   = Z_{RA} - comp 1

    A(27) = T_c, °K, comp 1

    A(28) = P_c, atm, comp 1

    A(29) = 273.16 °K, (0°C)

    A(30) = T_b, °K, comp 1*

    A(31) = T_c, °K, comp 2

    A(32) = P_c, atm, comp 2

    A( ) = T_b, °K, comp 2

    A( ) = acentric factor, comp 2**

    A( ) = Z_{RA} - comp 2*

    A(35) = molecular weight - comp 2

    A(36) = viscosity of comp 2*, centipoises

*   Apply to "A" only

** Apply to "B" only

Prediction of Diffusion Coefficients

Table 9.2
(Continued)

Program Notes:
1. For "A" physical data on cyclohexane(1) - benzene(2) system was used. For benzene, $\phi = 1.0$.
2. For the Scheibel correlation the "normal" value of K is given by equation 9.3. The exceptions are: 1) if water is the solvent and $V_1 < V_2$, then $K = 24.2 \times 10^{-8}$, 2) if benzene is the solvent and $V_1 < 2V_2$ then $K = 1.98 \times 10^{-8}$ and 3) for all other solvents where $V_1 < 2.5V_2$, $K = 17.5 \times 10^{-8}$.
3. For "B" the physical data of ethane-n-hexane were used in the example. For ethane $T_b = 184.5°K$ and for n-hexane, $341.9°K$.
4. $E = \varepsilon_1$ and ε_{12}, $F = \varepsilon_2$, $O = \sigma_1$ and σ_{12}, $Q = \sigma_2$, $I = \Omega_D$.
5. The constants of equation 9.6 are included in the program and are:
   A = 1.06036
   B = 0.1561
   C = 0.193
   D = 0.47635
   E = 1.03587
   F = 1.52996
   G = 1.76474
   H = 3.89411

equation 2.1. The latter may be calculated also, but in this program the viscosity of the solvent is entered before the calculations are started. Sample output, input and other information are given in Table 9.2.

9.2 DIFFUSION COEFFICIENTS FOR NON-POLAR GASEOUS MIXTURES

In Program 24, Table 9.1 and defined as "B" is the program of two correlations for the determination of values of the diffusion coefficients for mixtures of non-polar gases. In the gas phase, the diffusion coefficient is for practical purposes independent of composition, but is sensitive to pressure to some extent. The correlations programmed, the Chapman and Cowling (1961) and the Wilke Lee modification (1955) of this relationship include the pressure term ($D \propto 1/P$). This is valid up to moderate pressures. At high pressure the relationship is more complex.

The Chapman Cowling expression is:

$$D_{12} = 0.001858 T^{1.5} (((M_1 + M_2)/(M_1 M_2))^{.5}/P\sigma_{12}^2 \Omega_D) \qquad (9.5)$$

where Ω_D, the collision integral, may be expressed as:

$$\Omega_D = A/T^{*B} + C/\exp(DT^*) + E/\exp(FT^*) + G/\exp(HT^*) \qquad (9.6)$$

according to Neufeld, Janzen and Arig (1972).

$$T^* = kT/\varepsilon_{12} \qquad (9.7)$$

and

$$\varepsilon_{12} = (\varepsilon_1 \varepsilon_2)^{1/2} \qquad (9.8)$$

$$\sigma_{12} = (\sigma_1 + \sigma_2)/2 \qquad (9.9)$$

Tee, Gotoh and Stewart (1966) proposed correlations for length parameter, σ, and the energy parameter, ε--equations 7.13 and 7.14. A through H are empirical constants and their values are given in the program notes, Table 9.2.

In the Wilke-Lee modification of equation 9.5 the quantity-- $0.00217 - 0.00050 ((M_1 + M_2)/M_1 M_2)^{.5}$--is substituted for 0.001858.

Prediction of Diffusion Coefficients   101

To use the program for these two correlations, a value of the temperature (^{o}C) and pressure (atm) must be entered. Sample output, input data and other information is given in Table 9.2.

9.3 DIFFUSION COEFFICIENTS FOR POLAR GASEOUS MIXTURES

For mixture of polar gas, Brokaw (1969) recommended the use of the Chapman-Cowling relationship, equation 9.5, with the modification of the collision integral, Ω_D, as calculated by equation 9.6 modified as follows:

$$\Omega_D = \Omega_D \text{ (from eqn 9.6)} + (0.19\delta_{12}^2/T^*) \qquad (9.10)$$

where

$$\delta = ((1940\mu_p^2)/V_b T_b) \qquad (9.11)$$

$$\varepsilon/k = 1.18(1 + 1.3\delta^2)T_b \qquad (9.12)$$

$$\sigma = (1.585 V_b/(1 + 1.3\delta^2))^{1/3} \qquad (9.13)$$

$$\delta_{12} = (\delta_1 \delta_2)^{.5} \qquad (9.14)$$

$$\varepsilon_{12}/k = ((\varepsilon_1/k)(\varepsilon_2/k))^{.5} \qquad (9.15)$$

$$\sigma_{12} = (\sigma_1 \sigma_2)^{.5} \qquad (9.16)$$

Table 9.3
Program 25
Programs for Diffusion Coefficients of Mixtures of Polar Gases

```
1:"A"INPUT T,P
3:PRINT "T=";T
  ;"P=";P
5:T=T+A(29)
10:V=((R*A(27))
   /A(28))*Z^(1
   +((1-(A(30)/
   A(27)))^(2/7
   ))):D=(1940*
   G^2)/(V*A(30
   ))
13:A(37)=((R*A(
   31))/A(32))*
   A(34)^(1+((1
   -(A(33)/A(31
   )))^(2/7)))
15:A(39)=(1940*
   H^2)/(A(37)*
   A(33))
20:E=1.18*(1+(1
   .3*D^2))*A(3
   0)
22:F=1.18*(1+(1
   .3*A(39)^2))
   *A(33)
25:O=((1.585*V)
   /(1+(1.3*D^2
   )))^(1/3)
27:Q=((1.585*A(
   37))/(1+(1.3
   *A(39)^2)))^
   (1/3)
30:D=(D*A(39))^
   .5:O=(O*Q)^.
   5:E=(E*F)^.5
35:A(38)=T/E
40:I=(1.06036/(
   A(38)^.1561)
   )+(.193/EXP
   (.47635*A(38
   )))
43:I=I+(1.03587
   /EXP (1.5299
   6*A(38)))+(1
   .76474/EXP (
   3.89411*A(38
   )))
45:I=I+((.19*D^
   2)/A(38))
50:C=(.001858*T
   ^(3/2)*((M+A
   (35))/(M*A(3
   5)))^.5/(P*O
   ^2*I)
55:PRINT "D12="
   ;C;"--CM^2/S
   EC--BROKAW-M
   OD"
60:END
```

102

Prediction of Diffusion Coefficients 103

Table 9.4
Program 25
Sample Output, Input Data,
Storage Information, and Program Notes

```
T=15.14P=1.
D12=8.814671777E
-02--CM^2/SEC--B
ROKAW-MOD
T=64.34P=1.
D12=1.213912761E
-01--CM^2/SEC--B
ROKAW-MOD
```

Input Data: $t = 15.14°C$ $P = 1$ atm
$t = 64.34°C$ $P_2 = 1$ atm

Storage Information:

G   = dipole moment - comp 1

H   = dipole moment - comp 2

M   = molecular weight - comp 1

R   = 82.06, gas law constant, cm^3-atm/g mole $°K$

Z   = Z_{RA} - comp 1

A(27) = T_c, $°K$, comp 1

A(28) = P_c, atm, comp 2

A(29) = $273.16 °K$, ($0°C$)

A(30) = T_{b1}, $°K$, comp 1

A(31) = T_c, $°K$, comp 2

A(32) = P_c, atm, comp 2

A(33) = T_{b2}, $°K$, comp 2

A(34) = Z_{RA}, comp 2

A(35) = molecular weight - comp 2

Program Notes:

1. The physical data for the ammonia(1)-diethyl ether(2) were used in the example. For ammonia μ_p = 1.5 debyes and for diethyl-ether, 1.3.

2. In the calculated data E = ε_1 and ε_{12}, F = ε_2, O = σ_1 and σ_{21}, Q = σ_2, I = Ω_D, D = δ_1 and δ_{12}, A(39) = δ_2, V = V_{b1}, A(37) = V_{b2}, and A(38) = T/ε_{12}.

In the above expressions μ_p refers to the dipole moment and V_b the molal liquid volume at the normal boiling point, T_b. The value of V_b is calculated by the modified Rackett equation.

Program 25, Table 9.3, lists the Brokaw modification of the Chapman-Cowling relationship. The $t(^{\circ}C)$ and $P(atm)$ are the required input data. Sample output and other pertinent information are given in Table 9.4.

10
Surface Tension Correlations

A knowledge of the surface tension of pure liquids or liquid mixture is required in a number of process calculations. Perhaps the most common need occurs in the calculations of froth heights on trays in distillation columns and absorbers. In this chapter, programs are offered for the calculation of the surface tension of non-polar and polar pure liquids and for binary mixtures.

10.1 PURE LIQUIDS - NON-POLAR AND POLAR

Two correlations for the prediction of non-polar pure liquids have been programmed. The first is that of Brock and Bird (1955):

$$\alpha = P_c^{2/3} T_c^{1/3} Q(1 - T_r)^{11/9} \qquad (10.1)$$

where

$$Q = 0.1207 \left[1 + (T_{br}\ln P_c)/(1 - T_{br})\right] - 0.281 \qquad (10.2)$$

and the second is the Goldhammer (1910) modification of the Macleod (1923) - Sudgen (1924, a & b) equation:

$$\alpha = ([P]\rho_{Lb})^4((1 - T_r)/(1 - T_{br}))^{4n} \tag{10.3}$$

where $[P]$ is the parachor and the value of n depends on the type of compound. Values of n are given in the program notes. Program 26, Table 10.1, is a listing of the programs for equation 10.1 through

Table 10.1
Program 26
Programs for Calculation of Surface Tension of
Non-polar and Polar Liquids

```
1:"A"INPUT T                    45:"B"INPUT T
3:PRINT "T=";T                  47:PRINT "T=";T
5:T=T+A(29):A(                  50:T=T+A(29):A(
  40)=T/A(27):                     40)=T/A(27):
  A(41)=A(30)/                     A(41)=A(30)/
  A(27)                            A(27)
10:Q=(.1207*(1+                 55:Q=.1574+.359
  ((A(41)*LN (                     *W-1.769*X*X
  A(28)))/(1-A                     -.51*W^2+1.2
  (41)))))-.28                     89*W*X
  1                            60:M=1.21+.5385
15:O=A(28)^(2/3                    *W-14.61*X-3
  )*A(27)^(1/3                     2.07*X*X
  )*Q*(1-A(40)                  63:M=M-1.656*W^
  )^(11/9)                         2+22.03*W*X
20:PRINT "OMEGA                 65:O=A(28)^(2/3
  =";O;"--DYNE                     )*A(27)^(1/3
  S/CM--BROCK-                     )*Q*((1-A(40
  BIRD"                            ))/.4)^M
25:V=(R*A(27)/A                 70:PRINT "OMEGA
  (28))*Z^(1+(                     =";O;"--DYNE
  1-A(41))^(2/                     S/CM--HAKIM-
  7)):D=1/V                        ET-AL"
30:O=(C*D)^4*((                 75:V=(R*A(27)/A
  1-A(40))/(1-                     (28))*Z^(1+(
  A(41)))^(4*N                     1-A(41))^(2/
  )                                7)):D=1/V
35:PRINT "OMEGA                 80:O=(C*D)^4*((
  =";O;"--GOLD                     1-A(40))/(1-
  HAMMER-MOD-M                     A(41)))^(4*N
  ACLEOD-SUGDE                     )
  N"                           85:PRINT "OMEGA
40:END                             =";O;"--GOLD
                                   HAMMER-MOD-M
                                   ACLEOD-SUGDE
                                   N"
                                90:END
```

Surface Tension Correlations

Table 10.2
Program 26
Sample Output, Input Data,
Storage Information, and Program Notes

```
T=20.                      T=15.
OMEGA=28.0780125           OMEGA=17.3129575
7--DYNES/CM--BRO           9--DYNES/CM--HAK
CK-BIRD                    IM-ET-AL
OMEGA=28.0327184           OMEGA=17.8923618
3--GOLDHAMMER-MO           5--GOLDHAMMER-MO
D-MACLEOD-SUGDEN           D-MACLEOD-SUGDEN

       "A"                        "B"
```

Input Data: "A" t = 20°C
               "B" t = 15°C

Storage Information:

    C   = [P] parachor

    N   = exponent of term in equation 10.3*

    R   = gas law constant (82.06 cm^3 - atm/g mole °K)

    w   = acentric factor +

    X   = Stiel's polar factor +

    Z   = Z_{RA}

    A(27) = T_c, °K

    A(28) = P_c, atm

    A(29) = 273.16°K (0°C)

    A(30) = T_b, °K

Program Notes:

1. For "A" the physical data of benzene and for "B" those of diethyl ether were used in the examples.

               Benzene   Diethyl Ether

        n   0.29   ---

        X   ---   -0.003

* For "A" only

\+ For "B" only

10.3. The temperature is the only input variable, as surface tension is relatively insensitive to pressure. The liquid density at the normal boiling point, ρ_{Lb}, in equation 10.3 is calculated by the Spencer-Danner modification of the Rackett equation, 2.1. The program is defined as "A".

Sample output and other information are given in Table 10.2.

Also, included in program 26, defined as "B", are programs for the calculation of the surface tension of polar liquids. These are the correlations of Hakim et al. (1971) and equation 10.3 which can be used for polar substances as well as non-polar. Hakim et al. suggested the relationship:

$$\sigma = P_c^{2/3} T_c^{1/3} Q_p ((1-T_r)/.4)^m \tag{10.4}$$

where

$$Q_p = 0.1574 + 0.359\omega - 1.769X - 13.69X^2 - 0.510\omega^2 + 1.289\omega X \tag{10.5}$$

$$m = 1.210 + 0.5385\omega - 14.61X - 32.07X^2 - 1.656\omega^2 + 22.03\omega X \tag{10.6}$$

and X is Stiel's (1977) polar factor. Its value is given in the program notes. Sample output, input data, storage information and program notes are given in Table 10.2.

10.2 BINARY MIXTURES

The Macleod-Sudgen relationship (1923, 1924, a & b) when applied to mixtures becomes:

$$\sigma_m^{.25} = \sum_{i=1}^{n} [P_i](\rho_{Lm} x_i - \rho_{Vm} y_i) \tag{10.7}$$

Table 10.3
Problem 27
Program for the Calculation of
Surface Tension of Binary Mixtures

```
1:"A"INPUT T,P
  ,X,Y
3:PRINT "T=";T
  ;"P=";P;"X="
  ;X;"Y=";Y
5:T=T+A(29)
10:A(40)=(X*A(3
   0))/(X*A(30)
   +((1-X)*A(34
   )):A(41)=1-A
   (40)
15:A(42)=A(40)*
   A(27)+A(41)*
   A(32)
20:A(43)=X*A(31
   )+(1-X)*A(35
   )
25:A(50)=(X*A(2
   7)/A(28))+((
   1-X)*A(32)/A
   (33))
27:A(44)=R*A(50
   )*A(43)^(1+(
   1-(T/A(42)))
   ^(2/7))
30:D=1/A(44):Z=
   1
35:A=(.4278*A(2
   7)^2.5)/(A(2
   8)*T^2.5)
40:A(45)=(.4278
   *A(32)^2.5)/
   (A(33)*T^2.5
   )
42:B=(.0867*A(2
   7))/(A(28)*T
   )
45:A(46)=(.0867
   *A(32))/(A(3
   3)*T)
50:A=Y*A^.5+(1-
   Y)*A(45)^.5:
   B=Y*B+(1-Y)*
   A(46)
55:H=(B*P)/Z
60:A(47)=(1/(1-
   H))-(A^2*H/B
   )/(1+H)
65:IF ABS (Z-A(
   47))<.00001
   GOTO 100
70:Z=A(47)
75:PAUSE USING
   ;Z
80:GOTO 55
100:Z=A(47):A(48
    )=(Z*R*T)/P
105:O=(C*(X/A(44
    )-Y/A(48)))+
    (E*((1-X)/A(
    44)-(1-Y)/A(
    48)))
110:O=O^4
115:PRINT "OMEGA
    =";O;"--DYNE
    S/CM--MOD-MA
    CLEOD-SUGDEN
    "
120:END
```

Table 10.4
Program 27
Sample Output, Input Data,
Storage Information and Program Notes

```
T=65.P=63.62X=0.
233Y=0.438
OMEGA=2.43241479
3E-01--DYNES/CM-
-MOD-MACLEOD-SUG
DEN
T=-15.P=74.42X=0
.479Y=0.887
OMEGA=1.84220562
8--DYNES/CM--MOD
-MACLEOD-SUGDEN
```

Input Data: Run 1 t = 65. $°C$, P = 63.62 atm, x_1 = .233, y_1 = .438

Run 2 t = -15, P = 74.42, x_1 = 0.887

Storage Information:

C   = $[P]_1$, parachor, comp 1
E   = $[P]_2$, parachor, comp 2
R   = gas law constant (82.06 cm^3-atm/g mole $°K$)
A(27) = T_c, $°K$, comp 1
A(28) = P_c, atm, comp 1
A(29) = 273.16 $°K$, (0$°C$)
A(30) = V_{c1}, critical volume, cm^2/g mole, comp 1
A(31) = Z_{RA}, comp 1
A(32) = T_c, $°K$, comp 2
A(33) = P_c, atm, comp 2
A(34) = V_c, critical volume, cm^3/g mole, comp 2
A(35) = Z_{RA}, comp 2

Program Notes:

1. The physical data for the methane (1) - propane (2) system was used in example. For methane $[P]$ = 77 and for propane, 150.3.
2. Equilibrium compositions, temperature and pressure must be known to use program.
3. The calculated values of V_{1m} and V_{Vm} are stored in A(44) and A(48) respectively. The user may use other then calculated values. Steps 30-100 could then be eliminated and the values stored in the locations indicated.

Surface Tension Correlations

This equation has been programmed, Program 27, Table 10.3. The input data are the temperature, $^\circ$C, pressure, atm, and liquid and vapor phase mole fractions of component one, x_1 and y_1. In the program the liquid density, ρ_{Lm}, is determined by the Spencer and Danner modification of the Rackett equation 2.1. Li's method (eqns 2.16-17) is used to determine the critical temperature of the mixture and the vapor density is calculated by the Redligh-Kwong EOS, equation 1.1.

Sample output, input data, storage information, and program notes are included in Table 10.4.

11
Chemical Equilibrium Constants

Determination of chemical equilibrium constants are important since they permit the calculation of the theoretical degree of completeness of chemical reactions. While there are many equilibrium constants, the simplest one to calculate is that involving ideal gases as both reactants and resultants. Under these circumstances chemical equilibrium constants may be determined over a range of temperatures provided ideal gas state heat capacities of the various components as a function of temperature are known as well as the standard enthaply and entropy of the reaction.

11.1 EQUILIBRIUM REACTION CONSTANTS FOR IDEAL GAS

Using the reaction:

$$1/2 N_{2(g)} + 3/2 H_{2(g)} \rightarrow NH_{3(g)} \tag{11.1}$$

as a model, the equilibrium constant, K, assuming all components are in the ideal gas state, is defined as:

$$K = P_{NH_3} / P_{N_2}^{.5} P_{H_2}^{1.5} \tag{11.2}$$

Table 11.1
Program 28
Calculation of Reaction Equilibrium
Constant for Ideal Gases

```
1:"A"INPUT T
3:PRINT "T=";T
5:T=T+A(29)
10:A=A(40)-(.5*
   A(30)+1.5*A(
   35)):B=A(41)
   -(.5*A(31)+1
   .5*A(36))
15:C=A(42)-(.5*
   A(32)+1.5*A(
   37)):D=A(43)
   -(.5*A(33)+1
   .5*A(38))
20:I=H-(A*298.1
   6+(B*298.16^
   2)/2+(C*298.
   16^3)/3+(D*2
   98.16^4)/4)
25:J=S-(A*LN (2
   98.16)+B*298
   .16+(C*298.1
   6^2)/2+(D*29
   8.16^3)/3)
30:G=I/T-A*LN (
   T)-(B*T/2)-(
   C*T^2)/6-(D*
   T^3)/12+(A-J
   )
35:K=-(G/R):K=
   EXP (K)
40:PRINT "K=";K
   ,"H=";H;"S="
   ;S
45:END
```

The equilibrium constant is related to Gibbs free energy by:

$$-\Delta G^o/T = R\ln K \tag{11.3}$$

where

$$\Delta G^o = \Delta H^o - T\Delta S^o \tag{11.4}$$

Chemical Equilibrium Constants   115

Table 11.2
Program 28
Sample Output, Input Data,
Storage Information and Program Notes

T=526.84
K=3.029235574E-0
3H=-10960.S=-23.
6685

T=26.84
K=648.3633522H=-
10960.S=-23.6685

Input Data: Run 1   t = 526.84°C
              Run 2   t = 26.84°C

Storage Information:

H = ΔH°, standard enthalpy change for reaction, cal/g mole

R = gas law constant (1.987 cal/g mole °K)

S = ΔS°, standard entropy change for reaction, cal/g mole °K

A(29) = 273.16 °K (0°C)

A(30) through A(33) = heat capacity constants for reactant 1

A(35) through A(40) = heat capacity constants for reactant 2

A(40) through A(43) = heat capacity constants for resultant

Program Notes:

1. For the ammonia formation reaction: ΔH° = -10960 cal/g mole and ΔS° = -23.6685 cal/g mole °K.

2. Heat capacity constants for the substances in the ideal gas state are:

	N_2 (Reactant 1)	H_2 (Reactant 2)	NH_3 (Resultant)
a	6.903	6.952	6.5846
b	-3.753×10^{-4}	-4.576×10^{-4}	6.151×16^{-3}
c	1.93×10^{-6}	9.563×10^{-7}	2.3663×10^{-6}
d	-7.0×10^{-10}	-2.0×10^{-10}	-1.6×10^{-9}

3. Calculated values of I_H and I_S stored in I and J respectively Also Δa = A, Δb = B etc. $\Delta G^\circ/T$ = G

A convenient way to determine the value of ΔG^o as a function of temperature is through the relation:

$$\Delta G^o/T = I_H/T - \Delta a \ln T - \Delta b T/2 - \Delta c T^2/6$$
$$- \Delta d T^3/12 + (\Delta a - I_S) \tag{11.5}$$

where I_H and I_S are constants of integration and equal to:

$$I_H = \Delta H^o_{298} - (298.16 \Delta a + 298.16^2 \Delta b/2 + 298.16^3 \Delta c/3$$
$$+ 269.16^4 \Delta d/4) \tag{11.6}$$

$$I_S = \Delta S^o_{298} - (\ln 298.16 \Delta a + 298.16 \Delta b + 298.16^2 \Delta c/2$$
$$+ 298.16^3 \Delta d/3) \tag{11.7}$$

and Δa, Δb etc. are the differences in the empirical coefficients in the heat capacity expression:

$$C_p^* = a + bT + cT^2 + dT^3 \tag{11.8}$$

In the particular case under consideration, for example:

$$\Delta a = a_{NH_3} - (.5 a_{N_2} + 1.5 a_{H_2}) \tag{11.9}$$

$$\Delta b = b_{NH_3} - (.5 b_{N_2} + 1.5 b_{H_2}) \tag{11.10}$$

etc.

The basic expression is:

Chemical Equilibrium Constants

$$\Delta a = \Sigma(na)_{resultants} - \Sigma(na)_{reactants} \qquad (11.11)$$

By the above equations and a knowledge of the heat capacity expressions for the resultants and reactants, ΔH^o_{298}, and ΔS^o_{298} the value of the chemical equilibrium constant, K, may be determined at a given temperature. K values for the reaction for the production of ammonia from hydrogen and nitrogen (equation 1) can be determined by Program 28, Table 11.1. The input temperature, K, ΔH^o_{298} and ΔS^o_{298} are printed. Sample output, input data, and other information are given in Table 11.2.

12
Vapor Liquid Equilibrium

Vapor-liquid equilibrium (VLE) relationships are important in the design of distillation, columns, absorbers, partial condensers as well as other units. For this reason a great deal of effort has been devoted to reducing and correlating VLE data.

At sub-critical conditions, the liquid phase activity coefficient, γ, defined below, is usually employed for correlating VLE data.

$$\gamma = \hat{f}_i / x_i f^o_{Li} \tag{12.1}$$

where $\hat{f}_i$ is the fugacity of the component in the liquid mixture, and f^o_{Li} is the standard fugacity; i.e., the fugacity of pure component "i" in the liquid phase and at the temperature and pressure of the system. Under low pressure conditions, the activity coefficient may be approximated as the deviation from Raoult's law:

$$\gamma_i = y_i P / x_i P^o_i \tag{12.2}$$

Regardless how liquid phase activity coefficients are calculated, they are usually correlated by the Wilson (1964), Carlson-Colburn modification of the Van Laar (1942), or the Margules three-suffix (1895) equations. Each has two parameters which may be determined from infinite dilution activity coefficients or, in the special case of the formation of an azeotrope, from a knowledge of the azeotropic conditions.

While only binary mixtures will be treated here, all three correlating methods have been extended to multicomponent systems, with the Wilson equation being the most successful in this area.

12.1 PARAMETERS FROM γ^∞ VALUES

The expression for the activity coefficients in a binary mixture by the Wilson equation as modified by Orye and Prausnitz (1965) are:

$$\ln\gamma_1 = -\ln(x_1 + \Lambda_{12}x_2) + x_2(\Lambda_{12}/(x_1 + \Lambda_{12}x_2) \tag{12.3}$$
$$- \Lambda_{21}/(\Lambda_{21}x_1 + x_2))$$

$$\ln\gamma_2 = -\ln(x_2 + \Lambda_{21}x_1) - x_1(\Lambda_{12}/(x_1 + \Lambda_{12}x_2) \tag{12.4}$$
$$-\Lambda_{21}/(\Lambda_{21}x_1 + x_2))$$

at $x_1 = 0$, from equation 12.3:

$$\ln\gamma_1^\infty = -\ln\Lambda_{12} - \Lambda_{21} + 1 \tag{12.5}$$

and at $x_2 = 0$ from equation 12.4:

Vapor Liquid Equilibrium 121

$$\ln \gamma_2 = -\ln \Lambda_{21} - \Lambda_{12} + 1 \qquad (12.6)$$

where Λ_{12} and Λ_{21} are the Wilson Parameters.

If the values of the activity coefficients at infinite dilution are known, then the Wilson parameters may be determined using equations 12.5 and 12.6. Because of the logarithmic terms, a trial and error solution is required. The technique used in the program involves first making an assumption on the low side (e.g. .01) of the value of Λ_{12}, the value Λ_{21} is calculated by equations 12.5, followed by the calculation of γ_2^∞. When the calculated and given values of γ_2^∞ agree within an arbitrarily chosen limit ($\pm$ .002 in the program) satisfactory values of Λ_{12} and Λ_{21} have been obtained.

A difficulty in the trial and error procedures arises because the calculated value of γ_2^∞ changes very dramatically as the correct values of Wilson parameters are approached. Hence the trial and error procedure is set up to work only from the low side of the correct value of Λ_{12}. Initially the value of Λ_{12} is changed from 0.01 to 0.11 (i.e. Δ = 0.1) and γ_2^∞ calculated an outlined above. If the calculated value exceeds the given γ_{12}, Δ is subtracted from Λ_{12} and reduced by a factor of 10. The reduced Δ is added to Λ_{12}. This is step 35 in the program. Hence the correct value of Λ_{12} is always approached from the low side. On the other hand if after Λ_{12} is set at .11 the calculated value of γ_2^∞ remains below the given value, the Δ value is unchanged, step 30 in program. Once the calculated and given values of γ_2^∞ agree within the arbitrarily selected limits, the values of Λ_{12} and Λ_{12} and the calculated values of γ_2^∞ are printed.

The second correlation program used is the Marqules three suffix equation. The expression for the activity coefficients are:

$$\ln \gamma_1 = bx_2^2 + cx_2^3 \qquad (12.7)$$

$$\ln \gamma_2 = bx_1^2 + 3cx_1^2/2 - cx_1^3 \tag{12.8}$$

If $x_1 = 0$, equation 12.7 becomes:

$$\ln \gamma_1^\infty = b + c \tag{12.9}$$

and for $x_2 = 0$, equation 12.8 reduces to:

$$\ln \gamma_2^\infty = b + c/2 \tag{12.10}$$

The direct solution for the parameters b and c is possible and their values are printed out.

For the modified Van Laar equation:

$$\ln \gamma_1 = A/(1 + Ax_1/Bx_2)^2 \tag{12.11}$$

$$\ln \gamma_2 = B/(1 + Bx_2/Ax_1)^2 \tag{12.12}$$

hence

$$\ln \gamma_1 = A \tag{12.13}$$

$$\ln \gamma_2 = B \tag{12.14}$$

The values of A and B are printed in the program. The programs discussed above are included in Program 29, Table 12.1, and defined as "A". Sample output and other information is included in Table 12.2.

Vapor Liquid Equilibrium

Table 12.1
Program 29
Programs for the Calculation of Parameters of
Wilson, Van Laar, and Margales Equations

```
1:"A"INPUT I,J
3:PRINT "GAM1=
  ";I;"GAM2=";
  J
5:L=.01:X=A(27
  )
10:K=1-LN (L)-
   LN (I):M=-LN
   (K)-L+1:M=
   EXP (M)
12:PAUSE USING
   ;K
15:PAUSE USING
   ;M
20:D=M-J
25:IF ABS (D)<.
   002GOTO 100
30: IF D<0GOTO 7
   5
35:L=L-X:X=.1*X
   :L=L+X
40:GOTO 10
75:L=L+X
80:GOTO 10
100:PRINT "L12="
   ;L;"L21=";K;
   "GAM2=";M;"-
   -WILSON"
105:C=2*(LN (I)-
   LN (J)):B=LN
   (I)-C
110:PRINT "B=";B
   ;"C=";C;"--M
   ARGULES"
115:A=LN (I):E=
   LN (J)
120:PRINT "A=";A
   ;"B=";E;"--V
   AN-LAAR"
125:END

130:"B"INPUT X
132:PRINT "X1=";
   X
135:Y=1-X
140:G=-LN (X+L*Y
   )+(Y*((L/(X+
   L*Y))-(K/(K*
   X+Y))))
145:F=-LN (Y+K*X
   )-(X*((L/(X+
   L*Y))-(K/(K*
   X+Y))))
150:G=EXP (G):F=
   EXP (F)
155:PRINT "GAM1=
   ";G;"GAM2=";
   F;"--WILSON"
160:G=(A/(1+((A*
   X)/(E*Y)))^2
   ):F=(E/(1+((
   E*Y)/(A*X)))
   ^2)
165:G=EXP (G):F=
   EXP (F)
170:PRINT "GAM1=
   ";G;"GAM2=";
   F;"--VAN-LAA
   R"
175:G=B*Y^2+C*Y^
   3:F=(B+1.5*C
   )*X^2-C*X^3
180:G=EXP (G):F=
   EXP (F)
185:PRINT "GAM1=
   ";G;"GAM2=";
   F;"--MARGULE
   S"
190:END

200:"C"INPUT I,J
   ,X
203:PRINT "GAM1=
   ";I;"GAM2=";
   J;"X1=";X
205:Y=1-X:U=A(27
   ):L=.01
210:Q=LN (I)+LN
   (X+L*Y)-((L*
   Y)/(X+L*Y))
215:K=(-Q*Y/(Y+Q
   *X))
220:M=-LN (Y+K*X
   )-(X*((L/(X+
   L*Y))-(K/(K*
   X+Y)))):M=
   EXP (M)
225:PAUSE USING
   ;K
226:PAUSE USING
   ;M
230:D=M-J
235:IF ABS (D)<.
   0002GOTO 295
240:IF D>0GOTO 2
   70
245:L=L-U:U=U*.1
   :L=L+U
250:GOTO 210
270:L=L+U
275:GOTO 210
295:PRINT "L12="
   ;L;"L21=";K;
   "GAM2=";M;"-
   -WILSON"
300:A=LN (I)*(1+
   ((Y*LN (J))/
   (X*LN (I))))
   ^2
305:E=LN (J)*(1+
   ((X*LN (I))/
   (Y*LN (J))))
   ^2
310:PRINT "A=";A
   ;"B=";E;"--V
   AN-LAAR"
315:C=(Y^2*LN (J
   )-X^2*LN (I)
   )/(1.5*X^2*Y
   ^2-X^3*Y^2-Y
   ^3*X^2)
320:B=(LN (I)-C*
   Y^3)/Y^2
325:PRINT "B=";B
   ;"C=";C;"--M
   ARGULES"
330:END
```

Table 12.2
Program 29
Sample Output, Input Data,
Storage Information and Program Notes

GAM1=20.9894GAM2 =10.3039 L12=0.10206L21=0 .238176855GAM2=1 0.30554184--WILS ON B=1.62102738C=1. 422990168--MARGU LES A=3.044017548B=2 .332522464--VAN- LAAR	X1=0.3 GAM1=2.628945618 GAM2=1.309533219 --WILSON GAM1=3.49716601G AM2=1.34998424-- VAN-LAAR GAM1=3.605242528 GAM2=1.349285782 --MARGULES	GAM1=1.241GAM2=2 .047X1=0.61 L12=0.0967L21=4. 875596245E-01GAM 2=2.047064571--W ILSON A=2.10348335B=1. 551017804--VAN-L AAR B=1.02517166C=1. 011293122--MARGU LES
"A"	"B"	"C"

Input Data: "A" $\gamma_1^\infty = 20.9894$, $\gamma_2^\infty = 10.3039$

"B" $x_1 = 0.3$

"C" $\gamma_1 = 1.241$, $\gamma_2 = 2.047$, $x_1 = 0.61$

Storage Information:

A(27) = Value of Δ for "A" and "C"

Program Notes:

1. The original value of Δ has been set at 0.1 and need be entered only once. Also, its value may be changed by the user.
2. For "A" the VLE data ethanol-n-hexane system was used in the example. This system was also used in "B".
3. For "C" the azeotropic data for the methanol-benzene system was used in the example. The parameters calculated in this program may also be used in "B". The programs are compatible.
4. In "B" and "C" $x_2 = y$.

Vapor Liquid Equilibrium

12.2 PARAMETERS FROM AZEOTROPIC CONDITIONS

The parameters of the three correlations may also be determined from azeotropic data. For this unique condition, equation 12.1 becomes:

$$\gamma_i = P\hat{\phi}_i/f_{Li} \tag{12.15}$$

where $\hat{\phi}_i = \hat{f}_i/y_i P$ and equation 12.2 reduces to:

$$\gamma_i = P/P^S \tag{12.16}$$

In the case of the Wilson equation, equations 12.3 and 12.4 must be solved simultaneously. The technique used in similar to that described above. A low value is assumed for Λ_{12} (e.g. .01) and Λ_{21} calculated by equation 12.3. Then with values of Λ_{12} and Λ_{12}, a value of Λ_2 is calculated by equation 12.4. The calculated and given values of Λ_2 are compared and the correct solution obtained when these two values are in agreement. In the program a limit of ± .002 was arbitrarily selected. The same precaution discussed previously must be taken in this case, as well. That is, the correct value of Λ_{12} is approached from the low side and adjustments made in the size of the Δ quantity added to the assumed value of Λ_{12} are made so that this is the case. These are steps 240 and 245 in the program.

In the cases of the Margules and Van Laar equations, direct solution for the parameters are possible by equation 12.7, 12.8, 12.11 and 12.12. For the three correlations the parameters are printed. The program for the calculations are included in Program 29, Table 12.1, and defined as "C". Sample output, etc., is included in Table 12.2.

12.3 CALCULATIONS FOR GIVEN COMPOSITIONS

For completeness, included in Program 29 is a program which permits the calculations of the values of the liquid phase activity coefficients provided the liquid composition and the correlating parameters are known. In other words with the parameters of the Wilson, Margules, and Van Laar equation obtained through the use of either of the two programs described above, (or supplied by the user) the γ vs x relationships can be determined over the entire composition range ($x_1 = 0$ to $x_1 = 1.0$).

Equations 12.3, 12.4, 12.7, 12.8, 12.11 and 12.2 are used with a given input value of x_1. This program is included in Program 29, Table 12.1, and defined as "B". Sample output and other information is included in Table 12.2.

13
Fluid Friction and Orifice Calculations

Two common types of calculations in engineering design work are for the pressure drop in a flowing fluid due to friction and the sizing of orifices. There are several variations in each of these calculations depending upon the unknown or limiting factor (pressure drop, length, or quantity of flow, etc.) and type of fluid (incompressible or compressible). The following programs are for use in situations involving Newtonian fluid under isothermal conditions.

13.1 PRESSURE DROP CALCULATIONS/INCOMPRESSIBLE FLUIDS

The general equations used in the programs described below are:

$$\Delta P/\rho = (fLV^2)/(2Dg_c) \tag{13.1}$$

where f is the Moody fraction factor and equal to four times the Fanning friction factor. The friction factor - Reynolds number relationship proposed by Colebrook and White (1937) is:

$$1/\sqrt{f} = -2 \log((\varepsilon/D)/3.7 + 2.51/(Re\sqrt{f})) \tag{13.2}$$

where ε is the roughness and Re is the Reynolds number. In two special instances, simplifying assumptions are made:

1. For smooth pipes, (ε/D = 0), Colebrook and White suggested:

$$f = (1.8 \log (Re/7))^{-2} \qquad (13.3)$$

and
2. For streamline flow, Re < 2100, the Hagen-Poiseuille equation:

$$f = 64/Re \qquad (13.4)$$

is used.

Program 30, given in Table 13.1, makes use of the above equations and offers four types of solutions, each depending on the unknown variable. (ΔP, D, L, or Q) The program defined as "A" the pressure drop is the unknown, while, Q(gmp), h(ft), ε (symbolized by E, ft) and D(pipe diameter, inches) are known and are input data. Unless one of the two special cases is involved, a trial and error for the friction factor (eqn. 13.2) is required and accomplished in steps 605-652. The difference between the calculated value of f and and assumed value can be set by the user. After the value of f has been determined, the calculation is straight forward and the pressure drop, (psi), Re, and f are printed.

In the portion of the program defined as "B", the pipe diameter, D, is the unknown variable. The other variables are entered, followed by an estimated value of D(inches). The pressure drop is calculated and compared to the given value. (As in the previous solution the value of f is determined by trial and error). When the difference between the calculated pressure drop and the given one is less than .05 of the latter, a satisfactory solution is assumed. The differences can be set by the user. The results are printed.

Table 13.1
Program 30
Calculation of Pressure Drop, Pipe Diameter, Length
and Flow Rate for Incompressible Fluids

```
1:"A"INPUT Q,D          150:"C"INPUT Q,D        500:V=(Q*4)/(60*
  ,L,E                    ,P,E                        7.481*D^2*π)
2:PRINT "Q=";Q          152:PRINT "Q=";Q        510:A(40)=(D*V*0
  ;"-GPM";"--D             ;"--GPM";"D=                *1488)/M
  =";D;"-IN";"             ";D;"INCHES"         512:IF A(40)>210
  --L=";L;"-FT             ;"P=";P;"LBS              0GOTO 600
  ";"--E=";E;"             /IN^2";"E=";         513:IF A(40)<210
  -FT"                     E;"FT"                    0GOTO 685
3:F=.01                 153:F=.01               600:IF E=0GOTO 6
4:D=D/12                155:D=D/12                    90
5:GOSUB 500             157:PRINT "ESTIM        605:A=1/F^.5
20:PRINT "DEL-P             ATE-OF-LENGT        610:B=-2*LOG ((2
  =";A(30);"-L              H-FT"                     .51/(A(40)*F
  BS/IN^2";"RE          158:INPUT L                   ^.5))+(E/(D*
  =";A(40);"F=          160:GOSUB 500                 3.7)))
  ";F                   165:IF ABS (A(30       615:F=1/B^2
25:END                      )-P)<.05*P         618:IF ABS (A-B)
100:"B"INPUT Q,L            GOTO 190                 <.00002GOTO
  ,P,E                  170:IF A(30)>P                650
102:PRINT "Q=";Q            LET L=L-.1*L       619:PAUSE USING
  ;"-GPM";"L=";         172:IF A(30)<P                ;F
  ;L;"-FT";"P=                LET L=L+.1*L     630:GOTO 605
  ";P;"-LBS/IN          175:GOTO 160           650:F=(1/A)^2
  ^2";"E=";E;"          190:PRINT "DEL-P       652:GOTO 695
  -FT"                    =";A(30);"L=         685:F=64/A(40)
104:PRINT "ESTIM          ";L;"FT"             686:PAUSE USING
  ATE-OF-DIAME          195:END                      ;F
  TER-INCHES?"          200:"D"INPUT D,P       687:GOTO 695
105:INPUT D                 ,L,E               690:F=(1.8*LOG (
110:D=D/12              202:PRINT "D=";D              A(40)/7))^-2
112:F=.01                 ;"INCHES";"P         695:A(30)=(0*F*L
115:GOSUB 500             =";P;"LBS/IN               *V^2)/(2*144
120:IF ABS (A(30          ^2";"L=";L;"              *D*G)
  )-P)<.05*P              FT";"E=";E;"        700:RETURN
  GOTO 140                FT"
125:IF A(30)>P          203:D=D/12:F=.01
  LET D=D+.1*D          205:PRINT "ESTIM
128:IF A(30)<P              ATE-OF-Q-GPM
  LET D=D-.1*D              ?"
130:GOTO 115            206:INPUT Q
140:PRINT "DEL-P        210:GOSUB 500
  =";A(30);"LB          215:IF ABS (A(30
  S/IN^2";"F="              )-P)<.05*P
  ;F;"RE=";A(4              GOTO 240
  0)                    220:IF A(30)>P
141:D=D*12                  LET Q=Q-.1*Q
142:PRINT "D=";D        225:IF A(30)<P
145:END                     LET Q=Q+.1*Q
                        230:GOTO 210
                        240:PRINT "Q=";Q
                          ;"GPM"
                        245:END
```

Table 13.2
Program 30
Sample Output, Input Data,
Storage Information and Program Notes

```
Q=2.69-GPM--D=1.            Q=2.69--GPMD=1.0
049-IN--L=1000.-            49INCHESP=2.68LB
FT--E=0.00015-FT            S/IN^2E=0.00015F
DEL-P=2.68314741            T
9-LBS/IN^2RE=810            ESTIMATE-OF-LENG
4.846093F=3.4960            TH-FT
41089E-02                   DEL-P=2.59728670
                            2L=968.FT

        "A"                         "C"


Q=2.69-GPML=1000            D=1.049INCHESP=2
.-FTP=2.68-LBS/I            .68LBS/IN^2L=100
N^2E=0.00015-FT             0.FTE=0.00015FT
ESTIMATE-OF-DIAM            ESTIMATE-OF-Q-GP
ETER-INCHES?                M?
DEL-P=2.66426565            Q=2.7GPM
3LBS/IN^2F=3.495
758745E-02RE=809
3.537543
D=1.050465696

        "B"                         "D"
```

Inpute Data: "A"   Q = 2.69 gpm, D = 1.049 inches, L = 1000 ft,

E = 0.00015 ft

"B"   Q = 2.69 gpm, L = 1000 ft, dP = 2.68 psia,

E = 0.00015 ft

"C"   Q = 2.69 gpm, D = 1.049 inches, dP = 2.68 psi,

E = 0.00015 ft

"D"   D = 1.049 inches, dP = 2.68 psi, L = 1000 ft,

E = 0.00015 ft

Storage Data:

G = gravitational constant, 32.2 lb_f - ft/lb_m - sec^2

M = viscosity of fluid, cp

O = density of fluid, lbs/ft^3

Program Notes:

1. In "A" the pressure drop is unknown and f is obtained by trial and error.

2. In "B", an estimate of the diameter in inches is required.

Fluid Friction and Orifice Calculations 131

Table 13.2
(Continued)

Both the diameter and f must be obtained by trial and error.
3. In "C" an estimate of pipe length in feet is required. Both the length and f must be obtained by trial and error.
4. In "D" an estimate of the flow rate in gpm is required. The flow rate and f must be obtained by trial and error.
5. In the examples the physical properties of water at $70°F$ were used. Viscosity - 1 cp and density = 62.4 lbs/ft^3.

In the next portion, defined as "C", the length of the pipe, L, is the unknown variable. The same basic procedure is followed in this program as in the previous one; i.e., calculated and given values of the pressure drop as compared. In the last part, defined as "D", the flow rate, Q(gpm) is the unknown variable, and again the procedure described above is used. Sample output from the four programs and the other pertinent data are given in Table 13.2.

13.2 PRESSURE DROP CALCULATIONS FOR COMPRESSIBLE FLUIDS

The fact that the density of a compressible fluid varies with pressure must be taken into account when determining the pressure drop for this type of fluid. If the pressure drop is not too severe an average value of the density can be used. This was done in Program 31, Table 13.3. Also, the ideal gas law was used to calculate the densities. The user should be aware that sometimes an estimated value of the compressibility factor or a more complicated EOS may be required. The latter could be included as a sub-routine.

In Program 31, the pressure drop is the unknown variable and as a first guess P_2 is set equal to P_1 and the pressure drop calculated. The value of P_2 is then modified and the calculation repeated. Again the value of P_2 is calculated. A satisfactory solu-

Table 13.3
Program 31
Calculation of the Pressure Drop for Compressible Fluids

```
1:"A"INPUT Q
2:PRINT "Q=";Q
 ;"--FT^3/MIN
 "
5:D=D/12:T=T+A
 (29)
10:O=(P*N)/(R*T
 )
12:A(31)=0:A(32
 )=0:F=.01
15:GOSUB 100
20:IF ABS (A(33
 )-A(32))<.05
 *A(32)GOTO 5
 0
25:A(32)=A(33)
30:A(31)=(O+A(3
 2))/2
35:GOTO 15
50:PRINT "DEL-P
 =";A(30);"PS
 I"
51:D=D*12:T=T-A
 (29)
55:END
100:V=(Q*4)/(60*
 D^2*π)
105:U=(V*O)/A(31
 )
110:A(40)=(D*U*A
 (31)*1488)/M
115:IF A(40)>210
 0GOTO 150
120:IF A(40)<210
 0GOTO 180
150:IF E=0GOTO 2
 00
155:A=1/F^.5
160:B=-2*LOG ((2
 .51/(A(40)*F
 ^.5))+(E/(D*
 3.7)))
165:F=1/B^2
168:IF ABS (A-B)
 <.00002GOTO
 176
170:GOTO 155
176:F=(1/A)^2
178:GOTO 250
180:F=64/A(40)
185:GOTO 250
200:F=(1.8*LOG (
 A(40)/7))^2
250:A(30)=(A(31)
 *F*L*U^2)/(2
 *144*D*G)
255:A(35)=P-A(30
 )
257:A(33)=(A(35)
 *N)/(R*T)
259:PAUSE USING
 ;A(35)
265:RETURN
```

tion has been obtained when the values of P_2 agree within 5% (see step 20)--a restriction which is determined by the user. As is the case with incompressible fluid, the friction factor must be determined by trial and error. The results are printed. A sample output, input data and other information are given in Table 13.4.

When the pressure drop is one of the known variables, then an average value of the density can be calculated immediately. Hence, programs meeting this criterion described for imcompressible fluids may be used with a minor modification for compressible fluids. Therefore programs similar to those defined as "B", "C", and "D" in Table 13.1 are not repeated in Table 13.3.

Fluid Friction and Orifice Calculations

Table 13.4
Program 31
Sample Output, Input Data,
Storage Information and Program Notes

```
Q=1.8--FT^3/MIN
DEL-P=2.67444784
6E-01PSI
Q=18.--FT^3/MIN
DEL-P=25.7607602
6PSI
```

Input Data: Run 1   Q = 1.8 cfm

     Run 2   Q = 18 cfm

Storage Information:

  D = pipe diameter, inches

  E = roughness factor, ft (ε)

  G = gravitational constant, 32.3 ((lb_f - ft)/((lb_m - sec^2))

  L = length, ft

  M = viscosity, cp

  N - molecular weight

  P = initial pressure, psia

  R = gas law constant (10.73 psia - ft^3/lb mole - °R)

  T = temperature, °F

  A(29) = 460 °R (0°F)

Program Notes:

1. The physical data of air was used in the example and the ideal gas law was assumed. μ = .01807 cp at 70°F and molecular weight = 29.

2. Other data used were D = 1.049 inches, E = 0.00015 feet L = 1000 feet, P_1 = 50 psia, and T_1 = 70°F.

3. Calculated quantities include: F = friction factor, A(40) = Re, U = V_{avg}, and A(35) = P_2.

13.3 SIZING AN ORIFICE

The orifice is the most commonly used primary element in a system to measure the flow rate of a fluid. In sizing an orifice often the maximum flow rate and pressure drop are specified. Program 32, Table 13.5, can be used in this case. The basic equation used was (Bean (1971))

$$m = 0.52502 C Y D^2 \sqrt{\rho \Delta P}/\sqrt{1-\beta^4} \qquad (13.5)$$

where C is the coefficient of discharge, D the orifice diameter, β

Table 13.5
Program 32
Calculation of Orifice Diameters for
Maximum Flow and Pressure Drop

```
300:"A"INPUT Q,A
    (30)
302:PRINT "Q=";Q
    ;"--GPM--";"
    DEL-P=";A(30
    );"--PSI"
305:W=(Q*8.34)/6
    0
310:GOSUB 400
315:A(28)=A(27):
    B=A(27)/D
320:GOSUB 400
325:A(41)=A(27):
    B=A(27)/D
330:IF ABS (A(41
    )-A(28))<.01
    *A(41)GOTO 3
    50
335:GOTO 310
350:PRINT "ORIFI
    CE-DIAM=";A(
    28);"--IN"
355:END
400:A(27)=((W*(1
    -B^4)^.5)/(.
    52502*Y*C*(O
    *A(30))^.5))
    ^.5
402:PAUSE USING
    ;A(27)
405:RETURN

450:"B"INPUT Q,P
    ,A(30)
452:PRINT "Q=";Q
    ;"--CFM--";"
    P=";P;"-PSIA
    -";"DEL-P=";
    A(30);"-PSI"
453:Y=1:B=0
455:O=(P*M)/(R*(
    T+A(29))
460:W=(Q*O)/60
475:GOSUB 400
480:A(28)=A(27):
    B=A(27)/D
482:Y=1-(.41+.35
    *B^4)*(A(30)
    /(P*K))
485:GOSUB 400
490:A(41)=A(27):
    B=A(27)/D
492:Y=1-(.41+.35
    *B^4)*(A(30)
    /(P*K))
495:IF ABS (A(41
    )-A(28))<.00
    5*A(41)GOTO
    510
500:GOTO 475
510:PRINT "ORIFI
    CE-DIAM=";A(
    28);"--IN"
515:END
```

Fluid Friction and Orifice Calculations   135

Table 13.6
Program 31
Sample Output, Input Data,
Storage Information and Program Notes

```
Q=5.--GPM--DEL-P          Q=20.--CFM--P=50
=10.--PSI                 .-PSIA-DEL-P=5.-
ORIFICE-DIAM=2.8          PSI
88940178E-01--IN          ORIFICE-DIAM=4.5
                          23154738E-01--IN

        "A"                        "B"
```

Input Data: "A" Q = 5 gpm   dP = 10 psi

"B" Q = 20 cfm P_1 = 50 psia, dP = 5 psi

Storage Information:

    C   = discharge coefficient

    D   = pipe diameter, inches

    K   = ratio of C_p/C_v **

    M   = molecular weight **

    O   = density, lbs/ft^3 *

    R   = gas law constant (10.73 psi/1 mole °R)

    Y   = 1 *

    A(29) = 460°F (0°F)

Program Notes:

1. For "A" physical data of water at 70°F used in example. Density = 62.4 lbs/ft^3.
2. For "B" physical data of air at 70°F used in example. Molecular weight = 29 and K = 1.41.
3. In both example D = 0.622 inches, C = 0.62.

* For "A" only

** For "B" only

the ratio of diameters of orifice and pipe, and Y the expansion factor. Y = 1 for liquids and for gases, assuming the applicability of the ideal gas law, is:

$$Y = 1 - (0.41 + 0.35\beta^4)\Delta P/PK \tag{13.6}$$

where k is the ratio of heat capacities, Cp/Cv. The program defined as "A" is for the solution of 13.5 in the case the fluid is a liquid. The β term is unknown, since the diameter of the orifice is unknown, and initially set at zero. With this assumption the orifice diameter is estimated and then an iterative procedure is used until the estimated and calculated values of the orifice diameter agree within 1%. The results are printed. If the original estimate of the coefficient of discharge proves to be unsatisfactory, the value can be changed and the program re-run. A sample output, input data, and program notes are given in Table 13.6.

When sizing an orifice to measure the flow of gas, the Y factor (eqn. 13.6) is not unity, but a function of a number of factors including β, which initially are unknown. Thus an additional calculation is required and is the major difference between calculations for a liquid and those involving a gas. The program defined as "B" in Table 13.5 takes this difference into account. A sample output and other information are included in Table 13.6.

13.4 DETERMINATION OF LIQUID FLOWRATES

Equation 5 relating mass flowrate and pressure drop, can be written as:

$$m = CYA_o(2g_c\Delta\rho/(1 - \beta^4))^{.5} \tag{13.7}$$

Program 33, Table 13.7 offers the solution of the above equation. The program defined as "A" is for the case when the pressure drop (psi) is given and the flowrate, Q in GPM, is calculated. The

Fluid Friction and Orifice Calculations 137

Table 13.7
Program 33
Orifice Calculations for Flow Rate and
Pressure Drop for Noncompressible Fluids

```
 1:"A"INPUT A(3
   0)
 2:PRINT "DEL-P
   =";A(30);"--
   PSI"
 5:B=A(27)/D:Y=
   1
10:W=((C*Y*π*A(
   27)^2)/(4*14
   4))*((2*G*O*
   A(30)*144)/(
   1-B^4))^.5
11:PAUSE USING
   ;W
15:V=(W*4*144)/
   (O*π*D^2)
20:A(40)=(W*148
   8*4*12)/(π*A
   (27)*M)
25:A(41)=(W*60)
   /8.34
30:PRINT "GPM="
   ;A(41);"-REO
   =";A(40);"-V
   EL-PIPE=";V;
   "--FT/SEC"
35:END
40:"B"INPUT A(4
   1)
41:PRINT "GPM="
   ;A(41)
42:W=(A(41)*8.3
   4)/60
45:B=A(27)/D:Y=
   1
50:A(30)=((W*(1
   -B^4)^.5*4*1
   44)/(C*Y*π*A
   (27)^2*(2*G*
   O*144)^.5))^
   2
51:PAUSE USING
   ;A(30)
55:V=(W*4*144)/
   (O*π*D^2)
60:A(40)=(W*148
   8*4*12)/(π*A
   (27)*M)
65:PRINT "DEL-P
   =";A(30);"--
   PSI";"-REO="
   ;A(40);"-VEL
   -PIPE=";V;"F
   T/SEC"
```

Table 13.8
Program 33
Sample Output, Input Data,
Storage Information and Program Notes

```
DEL-P=5.--PSI              GPM=10.4518
GPM=10.45181296-           DEL-P=4.9999876-
REO=67269.66012-           -PSI-REO=67269.5
VEL-PIPE=3.87921           7668-VEL-PIPE=3.
1641--FT/SEC               879206829FT/SEC

        "A"                        "B"
```

Input Data: "A"   dP = r psia

                "B"   Q = 10.4518 gpm

Storage Information:

     C   = discharge coefficient

     D   = pipe diameter, inches

     G   = gravitational constant, 32.2 $(lb_f - ft)/(lb_m - sec^2)$

     M   = viscosity, cp

     O   = density, lbs/ft^3

     A(27) = orifice diameter

Program Notes:

1. The physical properties of water at $70°F$ were used in the example: viscosity = 0.932 cp, density = 62.4 lbs/ft^3.

2. Other data used were C = .61, D_p = 1.049 inches, D_o = 0.5 inches.

program defined as "B" is the reversed situation--the flowrate is given and the pressure drops the unknown. Sample output, input data, and program notes are included in Table 13.8.

13.5 DETERMINATION OF GAS FLOWRATES

Program 34, Table 13.9 is similar to the previous program with the exception that the effect of pressure on gas density must be taken into account and the expansion factor, Y, is not equal to unity,

Table 13.9
Program 34
Orifice Calculation for Flow Rate and
Pressure Drop for Compressible Fluids

```
70:"A"INPUT A(3
   0),P,T
71:PRINT "DEL-P
   =";A(30);"-P
   SI";"P1=";P;
   "-PSIA";"-T=
   ";T;"-F"
75:B=A(27)/D:Y=
   1-(.41+.35*B
   ^4)*(A(30)/(
   P*K))
80:O=(P*N)/(R*(
   T+A(29))
85:W=((C*Y*π*A(
   27)^2)/(4*14
   4))*((2*G*O*
   A(30)*144)/(
   1-B^4))^.5
86:PAUSE USING
   ;W
90:A(37)=(W*4*1
   44)/(π*D^2)
95:A(38)=(W*4*1
   44)/(π*A(27)
   ^2)
100:A(40)=(A(27)
    *1488*A(38))
    /(M*12)
101:PAUSE USING
    ;A(40)
105:Q=(W*60)/O
110:PRINT "Q=";Q
    ;"-CFM";"-G-
    PIPE=";A(37)
    ;"-LBS/SEC-F
    T^2";"-REO="
    ;A(40)
115:END

120:"B"INPUT Q,P
    ,T
121:PRINT "Q=";Q
    ;"-CFM";"P1=
    ";P;"-PSIA";
    "T=";T;"-F"
125:O=(P*N)/(R*(
    T+A(29)):B=A
    (27)/D:Y=1
130:W=(Q*O)/60
135:A(30)=((W*(1
    -B^4)^.5*4*1
    44)/(C*Y*π*A
    (27)^2*(2*G*
    O*144)^.5))^
    2
136:PAUSE USING
    ;A(30)
140:Y=1-(.41+.35
    *B^4)*(A(30)
    /(P*K))
145:X=((W*(1-B^4
    )^.5*4*144)/
    (C*Y*π*A(27)
    ^2*(2*G*O*14
    4)^.5))^2
150:IF ABS (X-A(
    30))<.01*A(3
    0)GOTO 170
155:Y=1-(.41+.35
    *B^4)*(A(30)
    /(P*K))
160:GOTO 135
170:A(37)=(W*4*1
    44)/(π*D^2)
175:A(38)=(W*4*1
    44)/(π*A(27)
    ^2)
180:PRINT "DEL-P
    =";A(30);"PS
    I";"-G-PIPE=
    ";A(37);"-LB
    S/SEC-FT^2";
    "REO=";A(40)
185:END
```

Table 13.10
Program 34
Sample Output, Input Data,
Storage Information and Program Notes

```
DEL-P=2.-PSIP1=5          Q=13.653-CFMP1=5
0.-PSIA-T=70.-F           0.-PSIAT=70.-F
Q=13.6546717-CF           DEL-P=1.99867346
M-G-PIPE=9.66731          PSI-G-PIPE=9.666
2221-LBS/SEC-FT^          981443-LBS/SEC-F
2-REQ=145837.709          T^2REQ=145837.70
?                         97
```

                              "A"   "B"

Input Data: For "A" dP = 2 psia, P_1 = 50 psia, E = 70°F

               For "B" Q = 13.653 cfm, P_1 = 50 psia, t = 70°F

Storage Information:

    C   = discharge coefficient

    D   = pipe diameter, inches

    G   = gravitational constants, 32.2 (lb_f - ft)/(lb_m - sec^2)

    K   = ratio of c_p/C_v

    M   = viscosity, c_p

    N   = molecular weight

    R   = gas law constant (10.73 psia - ft^3/lb mole - °R)

    A(27) = orifice diameter, inches

    A(29) = 460

Program Notes:

1. The physical properties of air were used in the example molecular weight = 29, viscosity = 0.01809 c_p and K = 1.41 and the ideal gas law was assumed valid.
2. The discharge coefficient = 0.61, Dp = 1.049 inches and D_o = 0.5 inches.
3. In "B", a trial and error solution was necessary. The expansion factor was initially set at unity.

Fluid Friction and Orifice Calculations   141

but calculated by equation 13.6. The program defined as "A" is the solution of the orifice equation (eqn. 11.7) for the flowrate (cfm) when the pressure drop and initial pressure and temperature are given. The program defined as "B" is the solution for the pressure drop with the flow rate given. Sample output, input data and program notes are given in Table 13.10.

14
Heat Transfer Calculations

Heat transfer calculations are common in engineering design work and are of many different types. In this chapter programs are given for the determination of heat transfer coefficients for fluids undergoing no phase change flowing inside and outside tubes and for vapors condensing on vertical and horizontal tubes. In addition, programs are given for the determination of the corrected mean temperature difference (CMTD) in a multipass shell and tube exchanger and the calculation of the overall heat transfer coefficient from individual heat transfer coefficients. This last item will be considered first.

14.1 CALCULATION OF THE OVERALL COEFFICIENTS

The relationship between the overall heat transfer and individual transfer coefficients is:

$$U_i = 1/(1/h_i + 1/h_{di} + x_w D_i/k_w D_{av} + D_i/D_o h_o + D_i/D_o h_{do}) \tag{14.1}$$

where x_w and k_w are wall thickness and thermal conductivity, respectively. The subscripts "i" and "o" refer to inside and outside

Table 14.1
Program 35
Calculation of Overall and Individual Heat Transfer
Coefficients. (h values for fluids without phase change)

```
1:"A"INPUT A(3
  1),A(32)
2:PRINT "HI=";
  A(31);"HO=";
  A(32);"--BTU
  /HR-FT^2-F"
3:IF A(30)=0
  LET A(30)=1.
  E10
4:IF A(33)=0
  LET A(33)=1.
  E10
5:U=(1/A(30)+1
  /A(31)+((O-I
  )*I)/(K*(O+I
  )*12)+(I/O)*
  (1/A(32)+1/A
  (33)))^-1
10:PRINT "U=";U
  ;"--BTU/HR-F
  T^2-F"
15:END

20:"B"INPUT Q
21:PRINT "Q=";Q
  ;"-GPM"
25:G=(Q*8.34*(D
  /62.4)*60*4*
  144)/(π*I^2)
30:A(40)=(I*G)/
  (12*M*2.42)
31:PAUSE USING
  ;A(40)
32:A(34)=M
33:Y=(1+(I/(12*
  L))^.7)
34:IF (L*12)/I>
  50LET Y=1
35:GOSUB 100
45:T=E-A(41)
51:PAUSE USING
  ;T
55:PRINT "VISCO
  SITY-AT-WALL
  -T?"
60:INPUT A(34)
65:GOSUB 100
66:A(43)=E-A(41
  )
68:IF ABS (T-A(
  43))<10GOTO
  80
70:T=A(43)
71:GOTO 51
80:PRINT "HI=";
  A(31);"--BTU
  /HR-FT^2-F";
  "HO=";A(32);
  "T-WALL=";T
85:END
100:A(31)=.023*(
  (K*12)/I)*Y*
  A(40)^.8*((C
  *M*2.42)/K)^
  (1/3)*(M/A(3
  4))^.14
101:PAUSE USING
  ;A(31)
105:A(41)=((1/A(
  31))/((1/A(3
  1))+(I/(O*A(
  32))))*(E-F)
110:RETURN
```

Heat Transfer Calculations   145

Table 14.1
(Continued)

```
200:"C"INPUT Q
201:PRINT "Q=";Q
    ;"-GPM"
203:X=J-O
205:G=(Q*8.34*(D
    /62.4)*60*4*
    144)/(π*(J^2
    -O^2))
210:A(40)=(X*G)/
    (12*M*2.42)
211:PAUSE USING
    ;A(40)
212:A(34)=M
213:Y=(1+(X/(12*
    L))^.7)
214:IF (L*12)/X>
    50LET Y=1
215:GOSUB 300
220:T=E-A(41)
221:PAUSE USING
    ;T
225:PRINT "VISCO
    SITY-AT-WALL
    -T?"
230:INPUT A(34)
235:GOSUB 300
240:A(43)=E-A(41
    )
245:IF ABS (T-A(
    43))<10GOTO
    280
250:T=A(43)
251:GOTO 221
280:PRINT "HO=";
    A(32);"--BTU
    /HR-FT^2-F";
    "HI=";A(31);
    "T-WALL=";T
285:END
300:A(32)=.023*(
    (K*12)/X)*Y*
    A(40)^.8*((C
    *M*2.42)/K)^
    (1/3)*(M/A(3
    4))^.14
301:PAUSE USING
    ;A(32)
305:A(41)=(1/A(3
    2)/((O/(I*A(
    31)))+1/A(32
    )))*(E-F)
310:RETURN

350:"D"INPUT Q,P
    ,T
351:PRINT "Q=";Q
    ;"--CFM";"-P
    1=";P;"--PSI
    A";"-T1=";T;
    "--F"
355:D=(P*N)/(R*(
    T+460))
360:G=(Q*4*144*D
    )/(60*π*I^2)
361:PAUSE USING
    ;G
365:A(31)=(.023*
    G^.8*K^(2/3)
    *C^(1/3))/((
    I/12)^.02*(M
    /1488)^.47)
370:PRINT "HI=";
    A(31);"BTU/H
    R-FT^2-F"
375:END
```

Table 14.2
Program 35
Sample Output, Input Data,
Storage Information and Program Notes

```
HI=200.H0=1000.-           Q=6.793-GPM
-BTU/HR-FT^2-F             VISCOSITY-AT-WAL
U=106.1844401--B           L-T?
TU/HR-FT^2-F               HI=383.1640295--
                           BTU/HR-FT^2-FH0=
                           1000.T-WALL=218.
                           1659641
HI=200.H0=1000.-
-BTU/HR-FT^2-F
U=157.9858416--B
TU/HR-FT^2-F

        "A"                        "B"


Q=6.8-GPM                  Q=10.--CFM-P1=50
VISCOSITY-AT-WAL           .--PSIA-T1=150.-
L-T?                       -F
H0=139.2627461--           HI=1.494816717BT
BTU/HR-FT^2-FHI=           U/HR-FT^2-F
400.T-WALL=149.4
806956

        "C"                        "D"
```

Input Data: For "A" h_i = 200 Btu/hr - ft^2 - $^\circ$F, h_o = 1000

For "B" Q = 6.793 gpm

For "C" Q = 6.8 gpm

For "D" W = 10 cpm, P_1 = 50 psia and t_1 = 150°F

Storage Information:

E = bulk temperature of cold fluid (benzene in example) +

F = bulk temperature of hot fluid +

I = outside diameter of pipe, inches

J = inside diameter of outside pipe (for "C")

K = thermal conductivity of pipe wall metal*

L = length of exchanger +

O = inside diameter of pipe, inches

Heat Transfer Calculations

Table 14.2
(Continued)

R = gas law constant (10.73 psia - ft^3/lb mole $^{\circ}$R)

$A(30)$ = h value for scale on inside pipe

$A(33)$ = h value for scale on outside pipe

Program Notes:
1. IN "A" for first run h_{di} = 500, ($A(30)$), and h_{do} = 800, ($A(33)$), and the second the two scale coefficients are set at zero. In the latter case, the program set the values of 10^{10} to avoid division by zero.
2. Pipe dimensions: i.d. = 2.067 inches and O.D. = 2.375 k_w = 26 Btu/hr - ft^2 - $^{\circ}$F (2" Sch. No 40 steel pipe).
3. For "B" and "C" physical properties of benzene at 110°F were used in example: μ = 0.48 C_p, ρ = 53.1 lbs/ft^3, k = 0.0898 Btu/hr - ft - $^{\circ}$F, C_p = 0.435 Btu/lb - $^{\circ}$F.
4. Tube dimensions I.D. = .745 inches, O.D. = 0.875 (copper), L = 150 ft, for "C" outer pipe I.D. = 1.610 inches.
5. For "B" t_w = 218 and μ_w = 0.23 cp. For "C" t_w = 149°F and μ_w = 0.36 cp.
6. For "B" bulk temperature of outside fluid = 250°F, h_o = 1000 Btu/hr - ft^2 - $^{\circ}$F. For "C" bulk temperature of inside fluid = 250°F and h_i = 400 Btu/hr - ft^2 - $^{\circ}$F.
7. For "D" physical properties of air at t = 150°F were used in the example k = 0.017 Btu/hr - ft - $^{\circ}$F, μ = .019 cp and C_p = 0.25 Btu/lb $^{\circ}$F.
8. Not all four programs can be loaded and used simultaneously.

* For "A" only. For other programs K = thermal conductivity of fluid.

+ For "B" and "C" only.

surfaces of the conduit separating two fluids. In Equation 14.1, the U value has been arbitrarily based on a square foot of area on the inside surface of the conduit.

In the program defined as "A" in Program 35, Table 14.1, the U_i value is calculated assuming all the individual h values are known. The h_i and h_o values are entered to start the calculations, while the other necessary data--conduit diameters, scale coefficients, etc. are in storage. To accomodate the case when one or both scale coefficients are zero, these quantities are set at 10^{10} in steps 3 and 4; thus avoiding division by zero. This program may be useful as a subroutine in larger programs. Sample output, input data, storage information and program notes are included in Table 14.2.

14.2 FLUIDS WITHOUT PHASE CHANGE

The program defined as "B" in Program 35, Table 14.1, is for the calculation of the heat transfer coefficient, h_i, for a fluid flowing in turbulent motion inside a circular conduit and without phase change. The relationship which includes the Sieder and Tate (1936) viscosity factor used is:

$$h_i D/k = 0.023 [1 + D/L)^{0.7}] (DG/\mu)^{0.8} (C_p \mu/k)^{1/3}$$
$$(\mu/\mu_w)^{0.14} \qquad (14.2)$$

Since the wall temperature is usually unknown, the viscosity ratio term is initially unknown and an iterative solution is necessary. The first approximation of h_i is made neglecting the (μ/μ_w) term. Then using that approximation the wall temperature is estimated by:

$$\Delta T_i = (1/h_i / (1/h_i + D_i/D_o h_o)) \Delta T \qquad (14.3)$$

and

Heat Transfer Calculations

$$T_w = T_{bi} - \Delta T_i \tag{14.4}$$

where ΔT is the difference between the bulk temperature of the fluid inside the conduit, T_{bi}, and the bulk temperature of the fluid outside the tube. Scale and wall resistance terms may be included in equation 14.3. To use equation 14.3 requires the value of the heat transfer coefficient, h_o, on the outside of the conduit. In the example it was assumed to be known.

After the wall temperature, T_w, has been calculated by equation 14.4 and the value shown to the user, step 51, the program calls for the fluid viscosity at T_w. The user enters the viscosity data and calculations proceed. The calculations are repeated and a second value of T_w determined. If this agrees within $\pm 10^o$ (difference is fixed by the user - step 68) of the first value, agreement is considered satisfactory and the results printed; otherwise another round of calculations follow. Sample output and other information are given in Table 14.2.

The program designated as "C" in Program 35, Table 14.2 is similar to that just described with the exception that the fluid is flowing in the annular space in a double pipe exchanger. In this case, the h_o value is calculated and the equivalent diameter, the difference between the inside diameter of the outside tube and the outside diameter of the inside tube, is used as the diameter term in equation 14.2. Sample output and pertinent information is included in Table 14.2.

The program defined as "D" in Program 35, Table 14.1 is for the solution of a simplified version of equation 14.2 which applied to gas and is:

$$h_i = 0.023(G^{0.8}k^{2/3}c_p^{1/3})/(D^{0.2}\mu^{0.47}) \tag{14.5}$$

Since the physical properties of gases are relatively insensitive to moderate changes in temperature, the viscosity ratio term may be neglected. Also the assumption was made that the D/L term is neglected. In Table 14.2, a sample output and other information are given. Equation 14.5 can be used for gases flowing outside tubes - in annular spaces, e.g. - and in that case the equivalent diameter is used.

14.3 CONDENSING VAPORS

In this section, programs are offered for the calculation of heat transfer coefficients for condensing on the outisde surfaces of vertical and horizontal tubes. The case of vertical tubes is considered in the program defined as "A", Program 36, Table 14.3. The relationship developed by Nusselt (1916):

$$h_o = 1.13((k_f^3 \rho_f^2 g \lambda)/(\Delta T_o L \mu_f))^{1/4} \tag{14.6}$$

was used.

In this relationship, the film temperature, T_f, must be determined by:

$$T_f = T_h + 3/4(\Delta T_o) \tag{14.7}$$

where $\Delta T_o = T_h - T_w$.

Hence to use equation 14.6, the wall temperature, T_w, and the temperature of the condensing vapor, T_h, are required. Presumably, the latter is known and the former temperature is calculated by equation 14.3 and 14.4, using an assumed value of h_o. This value is entered by the user to start the calculation. An estimated value of the T_f is calculated - step 15 - and shown. The user is then asked to enter values of k_f, ρ_f, μ_f - step 20. The value of h_o is recalculated and compared with the original estimate - step 35. If the two values agree within $\pm$ 5%, the results are printed.

Table 14.3
Program 36
Calculation of Heat Transfer
Coefficients for Condensing Vapors

```
1:"A"INPUT A(3
   5)
2:J=0
5:A(41)=1/A(35
   )/((O/(I*A(3
   1)))+((O-I)*
   I)/((O+I)*K*
   12)+1/A(35))
   *(E-F)
10:A(43)=E-A(41
   )
15:A(44)=F-.75*
   (F-A(43)):A(
   45)=F-A(43)
16:PRINT "TF=";
   A(44)
17:IF J>1.9GOTO
   30
20:PRINT "TC,RH
   O,MU-AT-TF?"
21:J=J+1
25:INPUT C,D,M
30:A(32)=1.13*(
   (C^3*D^2*G*3
   600^2*H)/(A(
   45)*L*M*2.42
   ))^.25
31:PAUSE USING
   ;A(32)
35:IF ABS (A(32
   )-A(35))<.05
   *A(32)GOTO 8
   0
40:A(35)=A(32)
45:GOTO 5
80:PRINT "HO=";
   A(32);"TF=";
   A(44);"TW=";
   A(43)
85:U=(1/A(31)+(
   (O-I)*I)/(K*
   (O+I)*12)+(I
   /(O*A(32))))
   ^-1
90:PRINT "UI=";
   U;"-BTU/HR-F
   T^2-F"
95:END

100:"B"INPUT A(3
    5)
101:J=0
105:A(41)=1/A(35
    )/((O/(I*A(3
    1)))+((O-I)*
    I)/((O+I)*K*
    12)+1/A(35))
    *(E-F)
110:A(43)=E-A(41
    )
115:A(44)=F-.75*
    (F-A(43)):A(
    45)=F-A(43)
116:PRINT "TF=";
    A(44)
117:IF J>1.9GOTO
    130
120:PRINT "TC,RH
    O,MU-AT-TF?"
121:J=J+1
125:INPUT C,D,M
130:A(32)=.725*(
    (C^3*D^2*G*3
    600^2*H)/(A(
    45)*N^(2/3)*
    (O/12)*M*2.4
    2))^.25
131:PAUSE USING
    ;A(32)
135:IF ABS (A(32
    )-A(35))<.05
    *A(32)GOTO 1
    80
140:A(35)=A(32)
145:GOTO 105
180:PRINT "HO=";
    A(32);"TF=";
    A(44);"TW=";
    A(43)
185:U=(1/A(31)+(
    (O-I)*I)/(K*
    (O+I)*12)+(I
    /(O*A(32))))
    ^-1
190:PRINT "UI=";
    U;"-BTU/HR-F
    T^2-F"
195:END
```

If the agreement between the calculated and estimated value is not satisfactory, the former is used as the estimated value and the calculations repeated. This iterative process is followed until a satisfactory agreement is reached. Each iteration yields a T_f and this, in turn, requires the user to enter values for the physical properties. Frequently after the second iteration the value of T_f is close to its final value, hence the property values change little during further trials. To reduce the number of times these values are required, a counter has been included in the program - see steps 2, 17, and 21. By this means, after the second values of k_f, ρ_f and μ_f are entered, the steps calling for these data are skipped. The user can increase this number by changing step 17. In addition, the h_o the U values are calculated and printed.

Sample output, input data and other information for this program are given in Table 14.4.

To calculate heat transfer coefficients of condensing vapors on horizontal tubes, the relationship (Nusselt 1916) used was:

$$h_o = 0.725((k_f^3 \rho_f^2 g \lambda)/(N^{2/3} \Delta T_o D_o \mu_f))^{1/4} \qquad (14.8)$$

where T_f is obtained as before - equation 14.7 - and N refers to the number of tubes in stack. The program defined as "B" in Program 36, Table 14.3 is for the solution of equation 14.8 and the method used is the same as that used in the case of vertical tubes. Sample output and other information is included in Table 14.4.

14.4 SHELL AND TUBE EXCHANGERS

In the simplest types of heat exchanger the shell side and tube side fluids make one pass through the exchanger and the flow of the two fluids is counter-current or co-current. In these cases a log mean temperature difference, LMTD, is used in the expression:

Heat Transfer Calculations

$$Q = UA \text{ LMTD} \tag{14.9}$$

In the majority of heat exchangers, one or both fluids make more than a single pass through the equipment. In these instances a corrected mean temperature difference (CMTD) is required and calculated from

$$\text{CMTD} = (\text{LMTD})F \tag{14.10}$$

The F factor is evaluated by relationships developed by Taborek (1951) and Gulley(1960) based on work by Bowman et al. (1940). These are:

$$F = \left[\frac{\sqrt{R^2+1}}{R-1}\right] \frac{\ln\left[(1 - P_x)/(1 - RP_x)\right]}{\ln\left[\frac{(2/P_x) - 1 - R + \sqrt{R^2+1}}{(2/P_x) - 1 - R - \sqrt{R^2+1}}\right]} \tag{14.11}$$

where

$$P_x = \frac{1 - \left[\frac{RP - 1}{P - 1}\right]^{1/N}}{R - \left[\frac{RP - 1}{P - 1}\right]^{1/N}} \tag{14.12}$$

and

$$P = (t_2 - t_1)/(T_1 - t_1) \tag{14.13}$$

$$R = (T_1 - T_2)/(t_2 - t_1) \tag{14.14}$$

See Figure 14.1 for explanation of terms.

Table 14.4
Program 36
Sample Output, Input Data,
Storage Information and Program Notes

```
     TF=204.6335408              TF=217.2380342
     TC,RHO,MU-AT-TF?            TC,RHO,MU-AT-TF?
     TF=221.6628292              TF=205.7128985
     TC,RHO,MU-AT-TF?            TC,RHO,MU-AT-TF?
     TF=219.7785795              TF=206.5749756
     HO=443.8607886TF            HO=881.9581129TF
     =219.7785795TW=1            =206.5749756TW=1
     93.038106                   75.4333008
     UI=145.6750164-B            UI=169.7617327-B
     TU/HR-FT^2-F                TU/HR-FT^2-F
```

       "A"   "B"

Input Data: For "A" Estimated value of h_o, Btu/hr - ft^2 - $°F$

       For "B" Estimated value of h_o, Btu/hr - ft^2 - $°F$

Storage Information:

  C = thermal conductivity of fluid, Btu/hr - ft - $°F$ +

  D = density of fluid, lbs/ft^3 +

  G = gravitational constant, 32.3 (lb$_f$ - ft)/(lb$_m$ - sec^2)

  H = latent heat of vaporization, Btu/lb

  I = inside diameter of pipe, inches

  K = thermal conductivity of metal pipe, Btu/hr - ft - $°F$

  M = fluid viscosity, cp +

  N = number of tubes in a stack *

  O = outside diameter of pipe, inches

  A(31) = heat transfer coefficient of fluid inside the pipe, h_i, Btu/hr - ft^2 - $°F$

Program Notes:

1. In both examples, steam at 300$°F$ was assumed as the condensing vapor.

2. The initial value of the film temperature depends upon the assumed value of h_o which is entered to start the program. The second value of the film temperature should normally be clase to the final value. Hence, only two values of k_f, p_f, and μ_f are requested. This is controlled by the counter "J". In "A" by steps 2, 17, and 21 and in "B" steps 101, 117, and 121. If additional changes in the film properties are necessary steps 17 and 117 can be altered.

3. In "B" N = 3

* "B" only

\+ Values are called for by the program after T_f is printed.

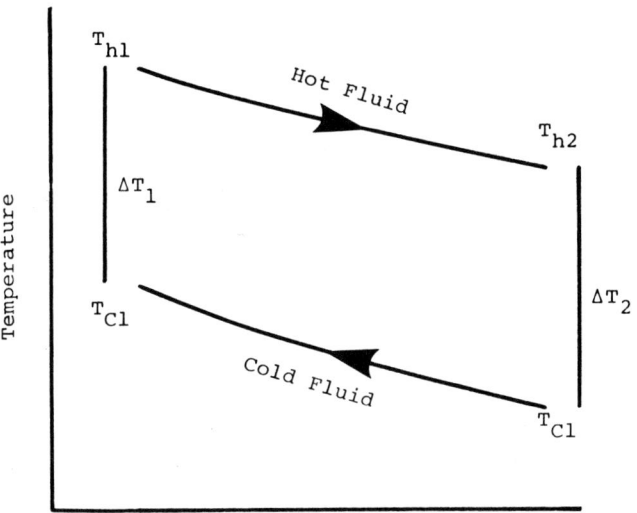

Figure 14.1

If $(T_{h1} - T_{h2}) = (T_{c2} - T_{c1})$, i.e., $R = 1$, then

$$F = \frac{P_x \sqrt{R^2+1}/(1-P_x)}{\ln\left[\dfrac{(2/P_x) - 1 - R + \sqrt{R^2+1}}{(2/P_x) - 1 - R - \sqrt{R^2+1}}\right]} \tag{14.15}$$

$$P_x = P/(N - NP + P) \tag{14.16}$$

where N is the number of passes on the shell side of the exchanger. Equations 14.11 through 14.16 are only valid for the simplest case

Table 14.5
Program 37
Calculation of Corrected Mean
Temperature Differences (CMTD)

```
1:"A"INPUT A,B
  ,C,D
2:PRINT "TH1="
  ;A;"TH2=";B;
  "TC1=";C;"TC
  2=";D
3:L=(A-D)-(B-C
  )
4:IF L=0GOTO 1
  0
5:L=((A-D)-(B-
  C))/LN ((A-D
  )/(B-C))
6:PRINT "LMTD=
  ";L
10:P=(D-C)/(A-C
   ):R=(A-B)/(D
   -C)
12:N=2
15:IF R=1GOTO 9
   0
20:X=(1-((R*P-1
   )/(P-1))^(1/
   N))/(R-((R*P
   -1)/(P-1))^(
   1/N))
25:F=((R^2+1)^.
   5/(R-1))*LN
   ((1-X)/(1-R*
   X))
30:F=F/LN (((2/
   X)-1-R+(R^2+
   1)^.5)/((2/X
   )-1-R-(R^2+1
   )^.5))
31:PAUSE USING
   ;F
35:IF F>.8GOTO
   60
40:N=N+1
45:GOTO 20
60:M=F*L
61:PRINT "CMTD=
   ";M;"F=";F;"
   N=";N
65:END
90:L=A-D
95:PRINT "LMTD=
   ";L
100:X=P/(N-N*P+P
    )
105:F=(X*(R^2+1)
    ^.5)/(1-X)
110:F=F/LN (((2/
    X)-1-R+(R^2+
    1)^.5)/((2/X
    )-1-R-(R^2+1
    )^.5))
111:PAUSE USING
    ;F
115:IF F>.8GOTO
    60
120:N=N+1
125:GOTO 100
```

Heat Transfer Calculations

Table 14.6
Program 37
Sample Output, Input Data,
Storage Information and Program Notes

```
TH1=250.TH2=100.
TC1=90.TC2=150.
LMTD=39.08650337
CMTD=34.19539291
F=8.748644664E-0
1N=3.
```

Input Data: $t_{h1} = 250°F$, $t_{h2} = 100°F$, $t_{c1} = 90°F$ and $t_{c2} = 150°F$

Storage Information: none

Program Notes:

1. L = (LMTD) and N = number of shell side passes and initially assumed as 2.

-- constant U and heat capacities, no change of phase takes place, and equal heat transfer surface in each pass, etc.

The program for the solution of the above equations is given in Program 37, Table 14.5 and defined as "A". In the program the value of P_x is determined for a given number of shell side passes, N. Usually a desired value of F is specified and this requires an iterative solution with changing values of N. Initially N = 2, step 12, and is increased by "1" with each iteration until the desired value of F is obtained. Sample output and other information are given in Table 14.6. The example given by Blackwell and Haydu (1981) was used.

15
Equilibrium Stage Calculations

In this chapter programs are given for the calculation of dewpoint and bubblepoint conditions, flash operations and the determination of the number of theoretical trays required in the stripping and rectifying sections of a distillation column. In addition, a program is offered for the determination of the number of trays required under conditions of total reflux. For these calculations the necessary equilibrium data were obtained either by Raoult's Law or the use of K(y/x) data. Programs for these methods were included as subroutines, so the user if free to use either one, or may substitute another. In the examples, two-component systems were used, but the methods are general and can be extended to multi-component systems with minor modifications.

15.1 DEWPOINT CALCULATIONS

Program 38, Table 15.1 lists the programs for three types of dewpoint calculations. In the program defined as "A" the values of y_1 (vapor phase mole fraction) and P (atm) are given and the calculation of T and x_1 (liquid phase mole fraction in equilibrium with vapor) are desired. To use the program the values of the known quantities are entered as well as are estimated value of the temperature (°C). A trial and error technique is used and Raoult's law is used to determine the y/x ratio. Where the Σx = 1.000 ± .0005 (step 14) the correct temperature has been determined. The value of .0005 may be changed by the user. The final temperature and x_1 values are printed.

Table 15.1
Program 38
Dewpoint Calculations

```
1:"A"INPUT Y,P
  ,A(10)
2:T=A(10)+273.
  16
3:GOSUB 100
4:PRINT "Y1=";
  Y;"P=";P
5:A(27)=(Y*P)/
  A(7)
6:A(28)=((1-Y)
  *P)/A(8)
7:A(29)=A(27)+
  A(28)
9:T=T+10
10:GOSUB 100
11:A(30)=(Y*P)/
  A(7)
12:A(31)=((1-Y)
  *P)/A(8)
13:A(32)=A(30)+
  A(31)
14:IF ABS (A(32
  )-1)<.0005
  GOTO 70
16:A(33)=T-((A(
  32)-1)*(T-A(
  10)-273.16))
  /(A(32)-A(29
  ))
18:T=A(33)
19:PAUSE USING
  ;T
21:GOTO 10
70:T=T-273.16
71:PAUSE USING
  ;T
72:PAUSE USING
  ;A(30)
73:PRINT "X1=";
  A(30);"T=";T
74:END
75:"B"INPUT T,P
76:T=T+273.16
77:GOSUB 100
78:A(27)=(1-(P/
  A(8)))/((P/A
  (7))-(P/A(8)
  ))
79:A(28)=A(27)*
  (P/A(7))
80:T=T-273.16
81:PRINT "T=";T
  ;"P=";P
82:PRINT "X1=";
  A(28);"Y1=";
  A(27)
83:END
```

Equilibrium Stage Calculations

Table 15.1
(Continued)

```
100:A(7)=A-(B/(T        110:"C"INPUT Y,T
   +C))                    ,A(40)
101:A(8)=D-(E/(T        111:PRINT "Y1=";
   +F))                    Y;"T=";T
102:A(7)=EXP (A(        112:T=T+273.16
   7))/760              114:GOSUB 100
103:A(8)=EXP (A(        120:X=(Y*A(40))/
   8))/760                 A(7)
104:RETURN              122:A(31)=((1-Y)
                           *A(40))/A(8)
                        125:Z=X+A(31)
                        130:A(40)=A(40)+
                           .5
                        135:A(32)=(Y*A(4
                           0))/A(7)
                        140:A(33)=((1-Y)
                           *A(40))/A(8)
                        145:Q=A(32)+A(33
                           )
                        150:IF ABS (Q-1)
                           <.0005GOTO 2
                           00
                        155:A(41)=A(40)-
                           ((Q-1)/((Q-Z
                           )/.5))
                        160:A(40)=A(41)
                        165:GOTO 135
                        200:T=T-273.16
                        205:PRINT "P=";A
                           (40);"X1=";A
                           (32)
                        210:END
```

Sample output and other required information for this program are given in Table 15.2.

In the program defined as "B" in Table 15.1, a knowledge of the temperature and pressure is assumed and the compositions of the equilibrium vapor and liquid are determined. This is a straight forward calculations and the results are printed. Table 15.2 includes a sample output and other information.

The program for third type of dewpoint calculation is defined as "C". In this case y_1 and T are known and x_1 and P are calculated. Again, this is a trial and error calculation and the correct pressure has been determined when the $\Sigma x = 1.000 \pm .0005$ (step 150).

Table 15.2
Program 38
Sample Output, Input Data,
Storage Information and Program Notes

Y1=0.25P=1.
X1=1.225332747E-
01T=105.1736432

T=105.17P=1.
X1=1.227965327E-
01Y1=2.505131631
E-01

Y1=0.25T=105.17
P=9.996510551E-0
1X1=0.122502229

"A" "B" "C"

Input Data: "A" $y_1 = 0.25$, $P = 1$ atm, and T_{est} °C
"B" $T = 105.17$ °C and $P = 1$ atm
"C" $y_1 = 0.25$, $t = 105.17$ °C, and P_{est}

Storage Information:

A
B   Antoine equation constants - comp 1
C
D
E   Antoine equation constants - comp 2
F

Program Notes:

1. The system benzene - toluene was used in the examples. The Antoine equation, eqn 3.6, was used to predict vapor pressure data. The following constants were used:

Antoine constant	Benzene	Toluene
A	15.9008	16.0137
B	2788.51	3096.52
C	-52.36	-53.67

Equilibrium Stage Calculations 163

The final results, x_1 and P are printed. Sample output is in Table 15.2.

In all of these programs Raoult's law was used to determine the y/x relationship. The programs can be expanded to multicomponent systems.

15.2 BUBBLEPOINT CALCULATIONS

Calculation of bubblepoint information is, in principle, the same as the previous dewpoint calculations. In this instance the liquid phase composition, x_1, is known and the vapor composition, y_1, is calculated. In Program 39, given in Table 15.3 the program defined as "A" applies to the case in which the temperature is unknown and the x_1 and pressure, P, are known. As before, an estimated temperature value is entered with the other input data and an iterative procedure is used. Where $\Sigma y = 1.0 \pm .0005$, the temperature value is satisfactory. The results are printed. In the program defined as "C", T and P are known and x_1 and y_1 calculated. Sample output and additional information for both of these programs are given in Table 15.4.

In these two programs the equilibrium data are used in the form:

$$\ln K = A + BT + CT^2 \tag{15.1}$$

An equation of this form is frequently adequate to express K(y/x) as a function of temperature under isobaric conditions. The program, defined as "B", allows the evaluate of the constants of equation 15.1 for both the components. Three sets of T, K_1 and K_2 are required and the values of constants for the two components are calculated (steps 100-130) and printed. The calculation of K for a given value of temperature is included as a subroutine - see steps

85 and 94. The equilibrium data must be known before the bubble-point programs can be used.

15.3 FLASH CALCULATIONS

There are a great variety of flash calculations, but the solution technique is general. To illustrate the procedure, programs are given for the solution of three types.

In the program defined as "A" in Program 40, Table 15.5, the known quantities are the fraction liquid product, L, the pressure, P, and the mole fraction of component one in the feed, x_{F1}. The

Table 15.3
Program 39
Bubblepoint Calculations

```
1:"A"INPUT X,A       50:"B"INPUT A(3
   (10)                 8)
2:PRINT "X1=";       55:A(30)=LN (A(
   X                    33)):A(31)=
5:T=A(10)+460           LN (A(34)):A
10:GOSUB 85             (32)=LN (A(3
12:Y=K*X:A(40)=         5))
   L*(1-X)           60:GOSUB 100
15:A(29)=Y+A(40      62:A=I:B=J:C=G
   )                 63:PRINT "A=";A
16:T=T+10               ;"B=";B;"C="
17:GOSUB 85             ;C
20:Y=K*X:A(40)=      64:A(30)=LN (A(
   L*(1-X)              36)):A(31)=
22:A(32)=Y+A(40         LN (A(37)):A
   )                    (32)=LN (A(3
25:IF ABS (A(32         8))
   )-1)<.0005        68:GOSUB 100
   GOTO 45           70:D=I:E=J:F=G
28:A(33)=T-((A(      71:PRINT "D=";D
   32)-1)*(T-A(         ;"E=";E;"F="
   10)-460))/(A         ;F
   (32)-A(29))      72:END
30:T=A(33)
31:PAUSE USING
   ;T
35:GOTO 17
45:T=T-460
46:PRINT "Y1=";
   Y;"T=";T
47:END
```

Table 15.3
(Continued)

```
85:K=A+(B*T)+(C
   *T^2)
87:K=EXP (K)
88:PAUSE USING
   ;K
90:L=D+(E*T)+(F
   *T^2)
92:L=EXP (L)
93:PAUSE USING
   ;L
94:RETURN
100:G=(((A(29)-A
   (27))*(A(31)
   -A(30)))-((A
   (32)-A(30))*
   (A(28)-A(27)
   )))
105:H=(((A(28)^2
   -A(27)^2)*(A
   (29)-A(27)))
   -((A(29)^2-A
   (27)^2)*(A(2
   8)-A(27))))
110:G=G/H
120:J=((A(31)-A(
   30))-(G*(A(2
   8)^2-A(27)^2
   )))/(A(28)-A
   (27))
125:I=A(30)-(J*A
   (27))-(G*A(2
   7)^2)
130:RETURN

150:"C"INPUT T
151:PRINT "T=";T
152:T=T+460
155:GOSUB 85
160:Y=(K-K/L)/(1
   -K/L)
161:X=Y/K
165:PRINT "Y1=";
   Y;"X1=";X
170:END
```

temperature and compositions of the product streams are unknown. The known quantities and an estimated temperature are entered and calculation proceed in iterative manner until $\Sigma y = 1.000 \pm .0005$ (step 30). When this criterion is satisfied, the correct temperature has been obtained. The results, T, x_1, y_1 and V are printed.

In the program defined as "B", Table 15.5, the liquid composition, x_1, pressure and feed composition, x_{f1}, are known and the temperature, vapor phase composition, y_1, and L are determined. The three known datum and an estimated value of the temperature are entered to start the calculation and the Newton technique used to converge on the correct answer. The test is made on the Σy's (step

Table 15.4
Program 31
Sample Output, Input Data,
Storage Information, and Program Notes

X1=0.4		T=128.08
Y1=0.8122823 T=12 8.0834141	A=-13.95301316 B=3.956697353E-02 C=-2.488790196E-05 D=-32.69005677 E=8.99953719E-02 F=-6.186373144E-05	Y1=8.119370838E-01 X1=3.998440551E-01
"A"	"B"	"C"

Input Data: "A" $x_1 = .40 \; T_{est} \; °F$
"B" $K_{23} = 1$ for comp 2 at t_3
"C" $t = 128.08 \; °F$

Storage Information:

$A(27) = T_1 - °R$
$A(28) = T_2 - °R$
$A(29) = T_3 - °R$
$A(33) = K_{11} - y/x$ ratio for comp 1 at T_1
$A(34) = K_{12} - y/x$ ratio for comp 1 at T_2
$A(35) = K_{13} - y/x$ ratio for comp 1 at T_3
$A(36) = K_{21} - y/x$ ratio for comp 2 at T_1
$A(37) = K_{22} - y/x$ ratio for comp 2 at T_2

---

228). In the flash calculation, defined as "C", x_{f1}, P and T are known and L, x_1 and y_1 are calculated by a trial and error procedure using the Newton Newton technique, step 330. An assumed L value is entered with the input data. The correct value of L is obtained when abs(Σx's - 1.0) is less than .00005 (step 325). The final results, L, x_1, and y_1 are printed. A subroutine for the calculation of the y/x ratio by Raoult's law is included in Program 40. Sample outputs and other information on these three programs are given in Table 15.6.

Equilibrium Stage Calculations

Table 15.4
(Continued)

A(38) = K_{23} - y/x ratio for comp 2 at T_3 - used at input data to start calculations.

Program Notes:

1. The propane-n-pentane system at P = 115 psia was used in the example. The K vs T data used were

T, °R	K_{C_3}	K_{n-C_5}
528	1.0	0.089
620	2.75	0.51
720	4.75	1.0

2. Calculation of A, B, C, equation 15.1, for the two component is necessary before programs "A" and "C" can be used.

3. The equation used to obtain values of A, B, and C were:

$$C = \frac{(T_3-T_1)(\ln K_{12} - \ln K_{11}) - (\ln K_{13} - \ln K_{11})(T_2 - T_1)}{(T_2^2 - T_1^2)(T_3 - T_1) - (T_3^2 - T_1^2)(T_2 - T_1)}$$

$$B = \frac{(\ln K_{12} - \ln K_{11}) - C(T_2^2 - T_1^2)}{(T_2 - T_1)}$$

$$C = \ln k_{11} - BT_1 - CT_1^2$$

4. For comp 2 D = A, E = B, and F = C.

15.4 THEORECTICAL PLATES - STRIPPING SECTION

The calculation of the theoretical plate below the feed plate (i.e. the stripping section) in a distillation column is a series of bubblepoint and material balance calculations. The first step is a bubblepoint calculation on the reboiler liquid - the bottom product. The next step is the use of a material balance to determine the

Table 15.5
Program 40
Flash Calculations

```
1:"A"INPUT L,P
  ,A(13),A(10)
2:PRINT "L=";L
  ;"P=";P;"XF1
  =";A(13)
3:A(14)=1-A(13
  ):V=1-L
4:T=A(10)+273.
  16
5:GOSUB 100
6:GOSUB 150
8: IF A(29)>1
  LET T=T-50
10:GOSUB 100
11:GOSUB 150
12: IF A(29)>1
  GOTO 160
15:I=10
17:T=T+I
18:PAUSE USING
  ;T
20:GOSUB 100
21:GOSUB 150
22: IF A(29)>1.0
  005LET T=T-I
23: IF A(29)>1.0
  005LET I=.1*
  I
25: IF A(29)>1.0
  005GOTO 17
30: IF ABS (A(29
  )-1)<.0005
  GOTO 50
35:GOTO 17
50:T=T-273.16
52:X=A(27)/A(7)
54:PRINT "T=";T
  ;"Y1=";A(27)
  ;"X1=";X;"V=
  ";V
55:END

200:"B"INPUT X,P
  ,A(13),A(10)
202:PRINT "X1=";
  X;"P=";P;"XF
  1=";A(13)
205:T=A(10)+273.
  16
207:GOSUB 100
210:GOSUB 120
212:A(29)=Y+A(28
  )
215:T=T+10
220:GOSUB 100
222:GOSUB 120
224:A(30)=Y+A(28
  )
225:PAUSE USING
  ;A(30)
228: IF ABS (A(30
  )-1)<.0005
  GOTO 250
230:A(11)=T-((A(
  30)-1)/((A(3
  0)-A(29))/(T
  -A(10)-273.1
  6)))
232:T=A(11)
235:PAUSE USING
  ;T
240:GOTO 220
250:T=T-273.16
255:L=(A(13)-Y)/
  (X-Y)
260:PRINT "X1=";
  X;"T=";T;"Y1
  =";Y;"L=";L
265:END
```

Table 15.5 (Continued)

```
300:"C"INPUT A(1           100:A(7)=A-(B/(T
    3),P,T,L                   +C))
302:PRINT "XF1="           101:A(8)=D-(E/(T
    ;A(13);"P=";               +F))
    P;"T=";T               102:A(7)=EXP (A(
303:T=T+273.16                 7))/760
305:GOSUB 100              103:A(8)=EXP (A(
308:A(40)=L                    8))/760
310:GOSUB 130              104:A(7)=A(7)/P
315:A(29)=X+A(28           105:A(8)=A(8)/P
    )                      106:RETURN
317:A(40)=L+.05            120:Y=A(7)*X:A(2
320:GOSUB 130                  8)=A(8)*(1-X
322:A(30)=X+A(28               )
    )                      125:RETURN
323:PAUSE USING            130:X=A(13)/(A(7
    ;A(30)                     )-A(40)*A(7)
325:IF ABS (A(30               +A(40))
    )-1)<.00005            135:A(28)=(1-A(1
    GOTO 350                   3))/(A(8)-A(
330:Z=A(40)-((A(               40)*A(8)+A(4
    30)-1)/((A(3               0))
    0)-A(29))/(A           137:RETURN
    (40)-L)))              150:A(27)=A(13)/
331:A(40)=Z                    ((L/A(7))+V)
332:PAUSE USING            153:A(28)=A(14)/
    ;A(40)                     ((L/A(8))+V)
335:GOTO 320               154:A(29)=A(27)+
350:Y=A(7)*X                   A(28)
352:PRINT "L=";A           156:RETURN
    (40);"X1=";X           160:PRINT "T-EST
    ;"Y1=";Y                   -TOO-HIGH-BY
355:END                        -50+,ENTER-N
                               EW-EST"
                           161:END
```

Table 15.6
Program 40
Sample Output, Input Data,
Storage Information and Program Notes

"A"	"B"	"C"
L=0.8P=1.XF1=0.2 5	X1=0.213P=1.XF1= 0.25	XF1=0.25P=1.T=10 1.6
T=101.6Y1=3.9612 97719E-01X1=2.13 4675571E-01V=0.2	X1=0.213T=101.58 80144Y1=3.951350 516E-01L=7.96854 0395E-01	L=7.963297943E-0 1X1=2.128966467E -01Y1=3.95070338 8E-01

Input Data: "A" $L = 0.8$, $P = 1$ atm, $x_{F_1} = 0.25$ and T_{est} $^\circ C$

"B" $x_1 = 0.213$, $P = 1$ atm, $x_{F_1} = 0.25$, and T_{est} $^\circ C$

"C" $x_{F_1} = 0.25$, $P = 1$ atm, $T = 101.6$ $^\circ C$, and L_{est}

Storage Information: See Table 15.2

Program Notes: See Table 15.2

Table 15.7
Program 41
Calculation of the Number of Theoretical
Trays in a Stripping Column

```
1:"A"INPUT A(2
  7),T
2:A(50)=((L-V)
  *A(27))/L
3:A(29)=T
4:N=0
5:T=T+273.16
7:A(28)=1-A(27
  )
9:GOSUB 100
10:A(29)=T-273.
  16
11:A(40)=A(7)+A
  (8)
12:PAUSE USING
  ;A(40)
13:T=T+10
15:GOSUB 100
17:Y=A(7)+A(8)
18:PAUSE USING
  ;Y
19:A(30)=T-((Y-
  1)/((Y-A(40)
  )/(T-A(29)-2
  73.16)))
21:IF ABS (Y-1)
  <.0005GOTO 5
  0
23:T=A(30)
24:PAUSE USING
  ;T
25:GOTO 15
50:T=T-273.16
51:PRINT "T=";T
  ;"Y1=";A(7);
  "X1=";A(27);
  "--PLATE-NO=
  ";N
53:IF T<94GOTO
  60
54:A(27)=((V*A(
  7))/L)+A(50)
55:T=T+273.16
56:N=N+1
57:GOTO 7
60:PRINT "L=";L
  ;"V=";V;"--M
  OLES/HR";"P=
  ";P
62:END
100:A(7)=A-(B/(C
   +T))
101:A(7)=EXP (A(
   7))/(760*P)
102:A(8)=D-(E/(F
   +T))
103:A(8)=EXP (A(
   8))/(760*P)
104:A(7)=A(7)*A(
   27)
105:A(8)=A(8)*A(
   28)
106:RETURN
```

Table 15.8
Program 41
Sample Output, Input Data,
Storage Information and Program Notes

```
T=109.5248364Y1=
5.366501588E-02X
1=0.0235--PLATE-
NO=0.
T=108.4468215Y1=
1.043558907E-01X
1=4.697830698E-0
2--PLATE-NO=1.
T=106.7115611Y1=
1.835728097E-01X
1=8.643248546E-0
2--PLATE-NO=2.
T=104.1444984Y1=
2.940650052E-01X
1=1.480893117E-0
1--PLATE-NO=3.
T=100.8273187Y1=
0.425462855X1=0.
234088593--PLATE
-NO=4.
T=97.2078919Y1=5
.538976363E-01X1
=3.363593515E-01
--PLATE-NO=5.
T=94.0178079Y1=6
.574249172E-01X1
=4.363238678E-01
--PLATE-NO=6.
T=91.6251804Y1=7
.277562929E-01X1
=0.516902154--PL
ATE-NO=7.
L=886.9V=690.3--
MOLES/HRP=1.
```

Input Data: x_{B1} = 0.0235 and T_{est} °C

Storage Information:

    A

    B   Antoine Equation Constants - Comp 1

    C

    D

    E   Antoine Equation Constants - Comp 2

    F

    L = liquid flow, moles/hr

    P = pressure, atm

    V = vapor flow, moles/hr

Program Notes:
1. The benzene-toluene system was used in the example. The Antoine constants for these compounds are given in Table 15.2.
2. P = 1 atm, k = 886.9 and V = 690.3 moles/hr. The feed plate temperature was set at 94°C.

composition of the liquid leaving the bottom tray. The material balance equation is:

$$x_1 = L_m - B/L_m + Bx_B/L_m \qquad (15.2)$$

This is followed by a bubblepoint calculation to determine the composition of vapor in equilibrium with the liquid on the 1st tray and the tray temperature. Equation 15.2 is then applied again, with the proper change in the subscripts, and so on until the feed tray is reached. Constant molal vapor and liquid flow were assumed (McCabe-Theile method) in the example.

By Program 41, Table 15.7, the number of plates is a stripping section of a distillation column can be determined. The mole fraction of the more volatile component in the bottoms product, x_B, is entered and the reboiler temperature and composition of the equilibrium vapor leaving the reboiler are calculated and printed. The plate number is also printed. For the reboiler this is "0". Next, equation 15.2 is applied - step 54 - and the composition of the liquid, x_1, on the tray above the reboiler determined. Then by a bubblepoint calculation the temperature and composition of equilibrium vapor are determined. x_1, y_1, and T and plate number are printed. The process is then repeated until the tray temperature exceeds the feed tray temperature, step 53. Then the calculation are stopped. The user can select a different criterion - the feed tray composition, e.g. - to stop the calculations.

In the example a two component system and Raoult's law were used. Both of these may be changed by the user without requiring major changes in the program. Sample output, input and other information are included in Table 15.8.

15.5 THEORETICAL PLATES - RECTIFYING SECTION AND TOTAL REFLUX

By the program defined as "A", Program 42, Table 15.9, the number of theoretical plates in the rectifying section of a distillation

Table 15.9
Program 42
Calculation of Number of Theoretical Trays
in a Rectifying Column and at Total Reflux

```
1:"A"INPUT A(2
    7),T
3:N=1
5:A(29)=1
10:A(50)=((V-L)
    *A(27))/V
15:T=T+273.16
20:A(28)=1-A(27
    )
25:GOTO 100
30:A(37)=A(27)
31:A(29)=T-273.
    16
35:A(38)=1-A(37
    )
40:GOSUB 200
45:GOSUB 270
50:A(60)=A(7)+A
    (8)
55:T=T+2
60:GOSUB 200
65:GOSUB 270
70:X=A(7)+A(8)
75:A(30)=T-((X-
    1)/((X-A(60)
    )/(T-A(29)-2
    73.16)))
80:IF ABS (X-1)
    <.0005GOTO 9
    0
82:T=A(30)
83:PAUSE USING
    ;T
85:GOTO 60
90:T=T-273.16
91:PRINT "T=";T
    ;"X1=";A(7);
    "Y1=";A(37);
    "-PLATE-NO="
    ;N
92:A(37)=((L*A(
    7))/V)+A(50)
93:N=N+1
94:IF T>96GOTO
    98
95:T=T+273.16
96:GOTO 31
98:PRINT "L=";L
    ;"V=";V;"--L
    B-MOLES/HR";
    "-P=";P
99:END

100:GOSUB 200
105:GOSUB 250
110:A(40)=A(7)+A
    (8)
115:T=T+10
120:GOSUB 200
125:GOSUB 250
130:Y=A(7)+A(8)
135:A(30)=T-((Y-
    1)/((Y-A(40)
    )/(T-A(29)-2
    73.16)))
140:IF ABS (Y-1)
    <.0005GOTO 1
    60
145:T=A(30)
146:PAUSE USING
    ;T
150:GOTO 120
160:T=T-273.16
161:PRINT "T=";T
    ;"XD=";A(27)
    ;"--CONDENSE
    R"
162:T=T+273.16
163:GOTO 30
200:A(7)=A-(B/(T
    +C))
205:A(7)=EXP (A(
    7))/(760*P)
210:A(8)=D-(E/(T
    +F))
215:A(8)=EXP (A(
    8))/(760*P)
220:RETURN
250:A(7)=A(7)*A(
    27)
255:A(8)=A(8)*A(
    28)
260:RETURN
270:A(7)=A(37)/A
    (7)
272:A(8)=A(38)/A
    (8)
275:RETURN

300:"B"INPUT A(3
    7),A(47),T
301:N=1
305:PRINT "XD1="
    ;A(37);"XB1=
    ";A(47)
310:A(29)=T
315:T=T+273.16
320:A(38)=1-A(37
    )
325:GOSUB 200
330:GOSUB 270
335:Z=A(7)+A(8)
340:T=T+5
345:GOSUB 200
350:GOSUB 270
355:X=A(7)+A(8)
360:A(30)=T-((X-
    1)/((X-Z)/(T
    -A(29)-273.1
    6)))
365:IF ABS (X-1)
    <.0005GOTO 4
    00
370:T=A(30)
375:GOTO 345
400:T=T-273.16
405:PRINT "T=";T
    ;"X1=";A(7);
    "Y1=";A(37);
    "--PLATE-NO=
    ";N
406:A(37)=A(7)
407:IF A(7)<A(47
    )GOTO 412
408:N=N+1
410:GOTO 310
412:END
```

column can be determined. These calculations are similar to those previously described. But, in this case a dewpoint calculation is used in conjunction with the material balance equation and a bubble-point calculation is used to calculate the condenser temperature if a total condenser is used.

In the program, the condenser temperature is calculated and printed after the composition of the distillation, x_D, has been entered. In this case (i.e., the use of a total condenser) the

Table 15.10
Program 42
Sample Output, Input Data,
Storage Information and Program Notes

"A"
```
T=80.6144442XD=0
.974--CONDENSER
T=81.4115164X1=9
.353717948E-01Y1
=0.974-PLATE-NO=
1.
T=82.8772956X1=8
.670316975E-01Y1
=9.439558404E-01
-PLATE-NO=2.
T=85.2620153X1=7
.617149577E-01Y1
=8.908024313E-01
-PLATE-NO=3.
T=88.6118301X1=6
.267034206E-01Y1
=8.088894115E-01
-PLATE-NO=4.
T=92.4304608X1=4
.886027348E-01Y1
=7.038804382E-01
-PLATE-NO=5.
T=95.9171311X1=3
.753891893E-01Y1
=5.964687937E-01
-PLATE-NO=6.
T=98.5242244X1=2
.977783003E-01Y1
=5.084138139E-01
-PLATE-NO=7.
L=536.9V=690.3--
LB-MOLES/HR-P=1.
```

"B"
```
XD1=0.974XB1=0.0
235
T=81.4090247X1=9
.354429571E-01Y1
=0.974--PLATE-NO
=1.
T=83.2643158X1=8
.492288016E-01Y1
=9.354429571E-01
--PLATE-NO=2.
T=87.0179139X1=6
.893895728E-01Y1
=8.492288016E-01
--PLATE-NO=3.
T=92.9216118X1=4
.719204675E-01Y1
=6.893895728E-01
--PLATE-NO=4.
T=99.5449423X1=2
.688193135E-01Y1
=4.719204675E-01
--PLATE-NO=5.
T=104.7363402X1=
1.332800503E-01Y
1=2.688193135E-0
1--PLATE-NO=6.
T=107.8256051X1=
6.096753096E-02Y
1=1.332800503E-0
1--PLATE-NO=7.
T=109.3657898X1=
2.680657176E-02Y
1=6.096753096E-0
2--PLATE-NO=8.
T=110.0710556X1=
1.157618702E-02Y
1=2.680657176E-0
2--PLATE-NO=9.
```

Equilibrium Stage Calculations   175

Table 15.10
(Continued)

Input Data: "A" x_{D1} = .974, T_{est} °C
              "B" x_{D1} = .974, x_{B1} = .0235, T_{est} °C

Storage Information:

   A
   B   Antoine constants for comp 1
   C
   D
   E   Antoine constants for comp 2
   F
   L = liquid flow, moles/hr*
   P = pressure, atm
   V = vapor flow, moles/hr*

Porgram Notes:
1. "A" cannot be run when "B" is in place, but "B" can be run when "A" is in place.
2. The benzene-toluene system at P = 1 atm was used in the example. Antoine constants for these compounds are given in Table 15.2.
3. For "A" L = 536.9 and V = 690.3 moles/hr.

* For "A" only.

---

composition of the vapor leaving the first (top) tray is the same as x_D and a dewpoint calculation is used to determine the tray temperature and the liquid phase composition, x_1. The results are printed as well as the tray number.

By the material balance:

$$y_{n+1} = L_m x_m / V_{m+1} + D x_D / V_{n+1} \qquad (15.3)$$

the composition of the vapor leaving the second tray is determined, step 92. Next the temperature and liquid composition on the second tray are determined by a dewpoint calculation. These results, y_1,

x_1, T and plate number, are printed. The process is then repeated--the application of equation 15.3 and a dewpoint calculation--until the tray temperature is below the feed tray temperature. Again, the user may select a different criterion to stop the calculation. As in the previous example, a two component system and Raoult's law were used. Sample output and other information are given in Table 15.10.

The program defined as "B" in Table 15.9 permits the calculation of the number of theoretical trays required at total reflux. This number is of interest because it is the minimum required for a given separation. At total reflux, the compositions of the liquid leaving and the vapors entering a given tray are identical-i.e., $y_{n+1} = x_n$. So, the calculations are simpler than those given in the two previous examples.

To use the program the distillate and bottoms compositions, x_D and x_B are entered. As in the previous example, the distillate composition is the same as the vapor composition leaving the top tray. The liquid composition and temperature on the top tray are determined by a dewpoint calculation. These results and the tray number are printed. Next, using the fact that $y_2 = x_1$, a second dewpoint calculation gives the liquid phase composition and temperature on the second tray. These results are printed and calculations continue until the calculated liquid phase composition equals, or is less rich in the more volatile component, than the specified bottoms composition.

Sample output and other program information are given in Table 15.10. As in most of the previous examples, a two component system and Raoult's law were used.

References

1. American Petrol. Inst., Research Project 44, Carnegie Press, (1953), p. 336.

2. Antoine, C., Compts. Redus., 107, No. 681, (1968), p. 836.

3. Bean, H.S., "Report of ASME Research Committee on Fluid Mechanics; 6th ed., Amer. Soc. of Mech. Engrs., (1971), p. 208.

4. Blackwell, W.W., and Haydu, L., Chem. Engr., Vol. 88, No. 17, (1981), p. 101-106.

5. Bondi, A., Ind. Engr. Chem. Funds., 5, (1966), p. 442.

6. Bowman, R.A., Mueller, A.C., and Hagle, W.M., Mean Temperature Difference in Design, Trans. ASME, Vol. 62, (1940), p. 283-294.

7. Brock, J.R., and Bird, R.B., AIChE J., 1, (1955), p. 174.

8. Brokaw, R.S., Ind. Engr. Chem. Proc. Des. & Dev., 8, No. 2, (1969), p. 240.

9. Carlson, H.C., and Colburn, A.P., Ind. Engr. Chem., 34, (1942), p. 581.

10. Carruth, G.F., and Kobayashi, R., Ind. Engr. Chem. Fund., 11, (1972), p. 509.

11. Chapman, S., and Cowling, T.G., "The Mathematical Theory of Nonuniform Gases," Cambridge, U. Press, New York, (1939).

12. Chapman, S., and Cowling, T.G., "The Mathematical Theory of Nonuniform Gases," Cambridge University Press, New York, (1961).

13. Cheuh, P.I., and Prausnitz, J.M., AIChE J., 13, (1967), p. 1099 and 15, (1969), p. 471.

14. Colebrook, C.F., and White, C.M., J. Inst. Civil Engrs., 10, No. 1, (1937-38), p. 471.

15. Dean, D.E., and Stiel, L.I., AIChE J., 11, (1965), p. 526.

16. Frost, A.A., and Kalkwarf, D.R., J. Chem. Phys., 21, (1953), p. 264.

17. Goldhammer, D.A., Z. Phys. Chem., 71, (1910), p. 577.

18. Gulley, D.L., Use Computers to Select Exchangers, Pet. Refiner. Vol. 39, No. 7, (1960).

19. Guggenheim, E.A., J. Chem. Phys., 13, (1945), p. 253.

20. Hakim, D.F., Steinberg, D., and Stiel, L.I., Ind. Engr. Chem. Fund., 10, (1971), p. 174.

21. Harlacher, E.A., and Braun, W.G., Ind. Engr. Chem. Proc. Des. Dev., 9, (1970), p. 479.

22. Jordan, H.B., M.S. Thesis, Louisiana State University, Baton Rouge, LA, (1961).

23. Jossi, J.A., Stiel, L.I., and Thodos, G., AIChE J., 8, (1962), p. 59.

24. Kay, W.B., Ind. Engr. Chem., 30, (1938), p. 459.

25. Kelley, K.K., U.S. Bur. Mines Bull., 584, (1960).

26. Lee, B.I., and Kesler, M.G., AIChE J., 21, (1975), p. 510.

27. Letsou, A., and Stiel, L.L., AIChE J., 19, (1973), p. 409.

28. Li, C.C., AIChE J., 22, (1976), p. 927.

29. Li, C.C., Canadian J. Chem. Engr., 19, (1971), p. 709.

30. Liley, P.E., Symposium on Thermal Properties, Purdue University, Lafayette, In, (1959), p. 40.

31. Lindsay, A.L., and Bromley, L.A., Ind. Engr. Chem., 42, (1950), p. 1508.

32. Lobe, V.M., M.S. Thesis, University of Rochester, Rochester, NY, (1973).

33. Lu, B.C.Y., Ruether, J.A., Hsi, C., and Chiu, D.H., J. Chem. Engr. Data, 18, (1973), p. 241.

REFERENCES

34. Lyckman, E.W., Eckert, C.A., and Prausnitz, J.M., Chem. Engr. Sci., 20, (1965), p. 703.

35. Macleod, D.B., Trans. Faraday Soc., 19, (1923), p. 38.

36. Margules, M., Sitzber, Akad. Wiss. Wien, Math, Maturw. Klasse II, 104, (1895), p. 1243.

37. Miller, D.G., reported in Reid, Prausnitz and Sherwood, "The Properties of Gases and Liquids," 3rd ed., (1977).

38. Missenard, A., Comptes Rendus, 260, No. 5, (1965), p. 5521.

39. Missenard, A., "Conductivite Thermique des Solides, Liquids Gas et De Leurs Melanges," Editions Eyrolles, Paris, (1965).

40. Newfeld, P.D., Janzen, A.R., and Aziz, R.A., J. Chem. Phys., 57, No. 3, (1972), p. 1105.

41. Neufeld, P.D., Janzen, A.R., and Aziz, R.A., J. Chem. Phys., 57, No. 3, (1972), p. 1100.

42. Nusselt, W., VDI Z. Vol. 60, (1916), p. 541, p. 569.

43. Orye, R.V. and Prausnitz, J.M., Ind. Engr. Chem., 57, (1965), p. 18.

44. Pitzer, K.S., Lippmann, D.A., Curl, R.F., Huggins, C.M., and Peterson, D.E., J. Am. Chem. Soc., 77, (1955), p. 3433.

45. Prausnitz, J.M., and Gunn, R.D., AIChE J., 4, (1958), pp 420 and 494.

46. Rackett, H.G., J. Chem. Engr. Data, 15, (1970), p. 514.

47. Rackett, H.G., J. Chem. Engr. Data, 16, (1971), p. 308.

48. Reddy, K.A., and Doraiswamy, L.K., Ind. Engr. Chem. Fund., 6, (1967), p. 77.

49. Reid, R.C., Prausnitz, J.M., and Sherwood, T.K., "The Properties of Gases and Liquids," 3rd ed., McGraw-Hill, New York (1977), p. 189 and 200.

50. Riedel, L., Chem. Ing. Tech., 23, (1949), p. 321.

51. Riedel, L., Chem. Ing. Tech., 26, (1951), pp. 59, 321, and 465.

52. Riedel, L., Chem. Ing. Tech., 26, (1954), p. 679.

53. Rowlinson, J.S., "Liquids and Liquid Mixtures," 2nd ed., Butterworths, London, (1969).

54. Scheibel, E.G., Ind. Engr. Chem., 46, (1954), p. 2007.

55. Sieder, E.M., and Tate, G.E., Ind. Engr. Chem., Vol. 28, (1936), pp. 14-29.

56. Spencer, C.F., and Danner, R.P., J. Chem. Engr. Data, 17, (1973), p. 230.

57. Spencer, C.P., and Danner, R.P., J. Chem. Engr. Data, 18, (1973), p. 230.

58. Spencer, C.F., and Adler, S.B., J. Chem. Engr. Data, 23, (1978), p. 92.

59. Stiel, L.I., Private Communication - See Reid, R.C., Prausnitz, J.M., and Sherwood, T.K., "Properties of Liquids and Gases", 3rd ed., McGraw-Hill Book Company, New York, (1977).

60. Stiel, L.I., and Thodos, G., AIChE J., 10, (1964), p. 274.

61. Sugden, S., J. Chem. Soc., (1924), p. 32.

62. Taborek, J.J., Organizing Heat Exchanger Programs on Digital Computers, Chem. Engr. Prog., Vol. 55, No. 10, (1959).

63. Tee, L.S., Gotoh, S., and Stewart, W.E., Ind. Engr. Chem. Fund., 5, (1966), pp. 356 and 363.

64. Thomas, L.H., J. Chem. Soc., (1946), p. 573.

65. Wassiljewa, A., Physik, 5, (1904), p. 737.

66. Watson, K.M., Ind. Engr. Chem., 23, (1943), p. 510.

67. Wilke, C.R., J. Chem. Phys., 18, (1950), p. 517.

68. Wilke, C.R., and Chang. P., AIChE J., 1, (1955), p. 264.

69. Wilke, C.R., and Lee, C.Y., Ind. Engr. Chem., 47, No. 6, (1955), p. 1253.

70. Wilson, G.M., J. Amer. Chem. Soc., 86, (1964), p. 127.

71. Yamada, T., and Gunn, J., J. Chem. Engr. Data, 18, (1973), p. 234.

72. Yoon, P., and Thodos, G., AIChE J., 16, (1970), p. 300.

Index

Activity coefficients, 119-126
　correlations, 120-124
　from azeotropic data, 125
　from infinite dilution data, 119-122
　from composition data, 125

Antoine vapor pressure equation, 41

Bondi equations for heat capacity, 41

Brock and Bird equation for surface tension of non-polar liquids, 105

Brokaw modification for polar gas phase diffusion coefficient, 101

Bubblepoint calculation, 163-164

Carlson and Colburn modification of van Laar equation for activity coefficients, 122, 125-126

Carruth and Kobayaski correlation for latent heat of vaporization, 47

Chapman and Cowling equation for non-polar gas phase diffusion coefficient, 100

Chapman and Enskog
　equation for viscosity on non-polar gases, 71
　equation to predict critical temperature of mixture, 32

Clapeyron equation for latent heat of vaporization, 47

Corrected mean temperature difference in multipass heat exchangers, 155-157

Critical temperatures of mixtures, 32-33

Dean and Stiel equation for effect of pressure on viscosity of a gas mixture, 78-81

Diffusion coefficients
　gas phase, 99-104
　liquid phase, 94-99

Dewpoint calculations, 159-163

Enthalpy (total) changes, 59-62

Entropy (total) changes, 59-62

Equilibrium stage calculations, 159-176
  bubblepoint, 163-164
  dewpoint, 159-163
  flash, 164-166
  theoretical trays, 167-176
    rectifying section, 172-176
    stripping section, 167-172
    total reflux, 176

Eucken equations for thermal conductivity of a gas, 88

Flash calculations, 164-166

Fluid friction, 127-133
  compressible fluids, 130-133
  incompressible fluids, 127-130

Friction factor correlations, 127-128

Frost-Kalkwarf-Thodos equation for vapor pressure, 44

Gas flow rates, measurement of, 130-141

Goldhammer modification for surface tension of non-polar liquids, 105

Gunn-Yamada equation for pure liquid volumes, 28

Hakim et al equation for surface tension of polar liquids, 108

Heat capacity of gases, 51-57
  isobaric enthalpy changes, 52, 56
  isobaric entropy changes, 52, 57

Heat transfer calculations, 143-157
  coefficients, 143-150
    condensing vapors, 150-153
    overall, 143
    fluids without phase change, 148-150

Isentropic process, 63-64

Jordan equation for thermal conductivity of a liquid mixture, 85

Jossi, Stiel and Thodos estimation of effect of pressure on viscosity of a gas, 72

$K(y/x)$ correlation, determination of constants, 163, 166-167

Kay's rule for prediction of critical temperature of a mixture, 32

Latent heat of vaporization, 45-49
  Carruth-Kobayashi, 47
  Clapeyron equation, 47
  Riedel equation, 45
  Watson equation, 47

Li
  method for prediction of critical temperature of a mixture, 32
  method to estimate thermal conductivity of a liquid mixture, 85

Liquid volumes, 27-33
  pure substances
    saturated, 27-29
    compressed, 30
  mixtures, 32-33

Liquid flow rates, measurement of, 136-138

INDEX 183

Lee Kesler
  correlation for vapor pressure, 37-40
  slope of vapor pressure-temperature curve, 49

Letsou and Stiel equation for viscosity of a pure liquid, 69

Lobe equation for viscosity of a binary liquid mixture, 69-71

Lu, Ruether, Hsi, and Chiu equation for volume of a pure liquid, 28

Lyckman, Eckert, and Prausnitz equation for volume of a pure liquid, 29

MacLeod-Sudgen modification for surface tension of a mixture, 108-111

Margules three suffix equation for activity coefficients, 121, 125-126

Maximum flow through an orifice, 134-136

Missenard equation for the estimation of thermal conductivity of a liquid, 84

Nusselt
  equation for heat transfer coefficients for condensing vapors, 150
    on horizontal tubes, 151
    on vertical tubes, 150

Orifice calculations, 134-141
  measurement of liquid flow rates, 136-138
  measurement of gas flow rates, 138-141
  sizing, 134-136

Peng-Robinson equation of state 15-26
pure substance, 15-26
  fugacity coefficient, 18
  multiple roots, 18
  thermodynamic departure functions, 18
  virial coefficients, 18

Pressure drop in flowing fluids, 127-133
  compressible fluids, 130-133
  incompressible fluids, 127-130

Reddy and Doraiswang equation for liquid phase diffusion coefficient, 96

Redlich-Kwong equation of state, 1-15
pure substance, 1-10
  fugacity coefficient, 6
  multiple roots, 6
  thermodynamic departure functions, 3
  virial coefficients, 6
mixtures, 10-15
  mixing rules, 10
  thermodynamic departure functions, 11

Riedel
  equation for latent heat of vaporization at normal boiling point, 45
  estimation of thermal conductivity of a liquid, 84

Riedel-Planck-Miller equation for vapor pressure, 41-44

Sato equation for estimation of thermal conductivity of a liquid, 83

Scheibel equation for liquid phase diffusion coefficient, 96

Sieder and Tate viscosity factor, 148

Spencer and Danner modification
  of Rackett equation
  for pure liquid volume, 27
  for liquid mixture voluem, 32

Stiel and Thodos
  equation for thermal conductivity of a gas, 88
  estimation of pressure effect on thermal conductivity, 88-91
  estimation of pressure effect on viscosity of a polar gas, 74-78

Surface tension, 105-111
  pure liquids, 105-111
  mixtures, 108-111

Tee, Botoh and Stewart equation for energy potential parameter and molecular diameter, 72

Thermal conductivity, 83-94
  gas, effect of pressure, 88-91
  pure, 88-91
  mixtures, 91-94
  liquid, pure, 83-85
  mixture, 85-87

Theoretical trays, calculation of number
  rectifying section, 172-176
  stripping section, 176-172
  total reflux, 176

Thomas equation for viscosity of a pure liquid, 67

van Laar equation, modified by Carlson and Colburn, for activity coefficients, 122, 125-126

Vapor-liquid equilibrium correlations, 119-126
  Carlson and Colburn modification of van Laar equation, 122, 125-126
  Margules, three suffix equation, 121-126, 126

(Vapor-liquid equilibrium correlations, 119-126)
  Wilson equation, 120, 125-126

Vapor pressure-temperature equations
  Antoine, 41
  Frost-Kalkwarf-Thodos, 44
  Lee-Kesler, 37-40
  Riedel-Planck-Miller, 41-44

Virial coefficients
  from Peng-Robinson EOS, 18
  from Redlich-Kwong EOS, 6

Viscosity
  gas, binary mixture, 78-81
    effect of pressure, 72-78
    non-polar, 71-75
    polar, 75-78
  liquid, binary mixture, 69-71
    pure, 32-35

Volume
  gas, pure, 1-3, 15-18
    mixture, 11
  liquid, pure, 27-32
    mixture, 32-35

Wassiljewa equation for thermal of a gas mixture, 91-92

Watson equation for latent heat of vaporization, 47

Wilke equation for the viscosity of a gas mixture, 78

Wilke and Chang, equation for liquid phase diffusion coefficient, 95

Wilke and Lee equation for gas phase diffusion coefficient, 100

Wilson correlation for activity coefficient, 119, 125-126

Yoon and Thodos equation for viscosity of a gas, 72